SEMA SIMAI Springer Series

Volume 24

As of 2013, the SIMAI Springer Series opens to SEMA in order to publish a joint series aiming to publish advanced textbooks, research-level monographs and collected works that focus on applications of mathematics to social and industrial problems, including biology, medicine, engineering, environment and finance. Mathematical and numerical modeling is playing a crucial role in the solution of the complex and interrelated problems faced nowadays not only by researchers operating in the field of basic sciences, but also in more directly applied and industrial sectors. This series is meant to host selected contributions focusing on the relevance of mathematics in real life applications and to provide useful reference material to students, academic and industrial researchers at an international level. Interdisciplinary contributions, showing a fruitful collaboration of mathematicians with researchers of other fields to address complex applications, are welcomed in this series.

THE SERIES IS INDEXED IN SCOPUS

More information about this series at http://www.springer.com/series/10532

David Greiner • María Isabel Asensio •
Rafael Montenegro

Editors

Numerical Simulation in Physics and Engineering: Trends and Applications

Lecture Notes of the XVIII 'Jacques-Louis Lions' Spanish-French School

 Springer

Editors
David Greiner
Instituto Universitario de Sistemas
Inteligentes y Aplicaciones Numéricas
en Ingeniería
Universidad de Las Palmas de Gran Canaria
Las Palmas de Gran Canaria, Spain

María Isabel Asensio
Departamento de Matemática Aplicada
Universidad de Salamanca
Salamanca, Spain

Rafael Montenegro
Instituto Universitario de Sistemas
Inteligentes y Aplicaciones Numéricas
en Ingeniería
Universidad de Las Palmas de Gran Canaria
Las Palmas de Gran Canaria, Spain

ISSN 2199-3041   ISSN 2199-305X   (electronic)
SEMA SIMAI Springer Series
ISBN 978-3-030-62545-0   ISBN 978-3-030-62543-6   (eBook)
https://doi.org/10.1007/978-3-030-62543-6

This Springer imprint is published by the registered company Springer Nature Switzerland AG.
The registered company address is: Gewerbestrasse 11, 6330 Cham, Switzerland

Preface

> *"One looks back with appreciation to the brilliant teachers,*
> *but with gratitude to those who touched our human feelings.*
> *The curriculum is so much necessary raw material,*
> *but warmth is the vital element for the growing plant*
> *and for the soul of the child."*
>
> Carl Jung

This book contains the lecture notes of the XVIII Spanish-French School "Jacques Louis Lions" on Numerical Simulation in Physics and Engineering (http://ehf2018. iusiani.ulpgc.es), which took place from 25th to 29th June 2018 in the University of Las Palmas de Gran Canaria (ULPGC), organized by the Institute of Intelligent Systems and Numerical Applications in Engineering (SIANI), the Department of Mathematics, and the Department of Civil Engineering of ULPGC, and the Research Group in Numerical Simulation and Scientific Calculus (SINUMCC) of University of Salamanca.

The Spanish-French Schools on Numerical Simulation in Physics and Engineering are held biennially since 1984, becoming a meeting point for professionals, researchers, and students in the field of numerical methods. These conferences are sponsored by the Spanish Society of Applied Mathematics (SEMA), actively involved in the organization of these schools that, together with the Congresses of Differential Equations and Applications (CEDYA)/Congresses of Applied Mathematics (CMA), constitute the two most important series of scientific meetings sponsored. They also have the sponsorship of the Société de Mathématiques Appliquées et Industrielles (SMAI) of France since 2008. The 17 previous editions were held in Santiago de Compostela (1984), Benalmádena (1986), Madrid (1988), Santiago de Compostela (1990), Benicàssim (1992), Sevilla (1994), Oviedo (1996), Córdoba (1998), Laredo (2000), Jaca (2002), Cádiz (2004), Castro Urdiales (2006), Valladolid (2008), A Coruña (2010), Torremolinos (2012), Pamplona (2014), and Gijón (2016).

Each edition is organized around several main courses delivered by renowned French and Spanish scientists. On this last occasion, there were four 4-h courses, for which the lecturers were Philippe Destuynder, Héctor Gómez, Frederic Hecht, and José Sarrate, together with three 1-h talks by Luis Ferragut, Raphaele Herbin, and Jacques Periaux. In addition, four 30-minute talks were given by: SeMA "Antonio Valle" Young Investigator Award 2017, Javier Gómez Serrano, best thesis 2017 selected for ECCOMAS award, Adrián Moure, best paper 2016 of SeMA Journal, Serge Nicaise, and best paper 2017 of SeMA Journal, Pablo Pedregal. Furthermore, the participants in the School had the opportunity to present their research work in two poster sessions, where a total of 24 posters were exposed. Finally, a series of ten brief lectures were given by some of the former students, or direct scientific disciples of them, of Professor Luis Ferragut Canals (University of Salamanca), who himself acted as chairman of the session, in recognition of its special influence in the formation of applied mathematics groups in Las Palmas de Gran Canaria, Madrid, and Salamanca.

The editors warmly thank all the speakers and participants for their contributions to the success of the School. In particular, we would like to acknowledge the efforts of all the lecturers (first and second chapters) and speakers (third and fourth chapters) who have contributed to this SEMA SIMAI Springer Series volume, which additionally contains the two "ex-aequo" awarded posters (fifth and sixth chapters). We are also grateful to the Organizing and Scientific Committees for their efforts in the preparation of the School. We extend our thanks and gratitude to all sponsors and supporting institutions for their valuable contributions: SEMA, SMAI, University of Las Palmas de Gran Canaria, and their Departments of Mathematics and Civil Engineering, as well as the Institute of Intelligent Systems and Numerical Applications in Engineering (SIANI). Finally, the editors acknowledge SEMA SIMAI Springer Series and their editors-in-chief Luca Formaggia and Pablo Pedregal for the interest to this series in publishing the most representative scientific and industrial material presented in the Spanish-French School 'Jacques Louis Lions'. Since its XV edition, these lecture notes have been continuously published, this being the fourth volume held in this series.

Las Palmas de Gran Canaria, Spain   David Greiner
Salamanca, Spain   María Isabel Asensio
Las Palmas de Gran Canaria, Spain   Rafael Montenegro
September 2020

Contents

About the Editors

David Greiner is Associate Professor at the Institute of Intelligent Systems and Numerical Applications in Engineering of University of Las Palmas de Gran Canaria (ULPGC) and has served as General Chair of the XVIII Spanish-French School "Jacques-Louis Lions" on Numerical Simulation in Physics and Engineering. His main research interests are related to optimum design in engineering. He is also editorial board member for the Mathematics—MDPI and Mathematical Problems in Engineering journals.

María Isabel Asensio received her PhD in Mathematics from the University of Salamanca, where she is Associate Professor in the Department of Applied Mathematics and Principal Investigator of the Numerical Simulations and Scientific Computation Research Group. Her main research interests concern mathematical modelling of environmental problems with emphasis on forest fire spread, wind field, and atmospheric pollutants' dispersion.

Rafael Montenegro is Professor of Applied Mathematics at the University of Las Palmas de Gran Canaria (ULPGC). He was Director of the Mathematic Department and the University Institute SIANI. He is editorial board member of relevant journals and is the author of more than 200 scientific publications, mainly in the area of 3-D adaptive finite element mesh generation and its applications to environmental problems. He is currently member of the Managing Board of ECCOMAS in representation of SeMA.

An Introduction to Quasi-Static Aeroelasticity

Philippe Destuynder and Caroline Fabre

Abstract The aeroelasticity is the science which models, analyses and describes the coupled movements between a flow and a flexible structure. The different phenomena encountered can be classified using three (at least) adimensional numbers: the Strouhal number, the Reynolds number and the reduce frequency number (which despite its name, has no dimension). For sake of clarity, let us just mention in this abstract, that the reduce frequency is the ratio between the time necessary to a flow particle for flying over a flexible structure and the fundamental period of oscillation of this structure.

In the framework of quasi-static aeroelasticity, it is always assumed that the reduce frequency is smaller than the unity. It enables one to define the flow fields (velocity, pressure) from a static position of the structure. The effect of its position with respect to the flow leads to a modification of the stiffness (added aerodynamic stiffness). Furthermore, the relative flow velocity (difference between the flow velocity and the one of the structure) leads to introduce damping due to the flow and therefore modifies the static analysis of stability into the dynamic stability study (aerodynamic damping).

Recently, engineers have upgraded this approach by introducing the added mass concept. This is a mechanical effect due to the fact that the inertia of the structure should take into account the mass of flow which is involved in a movement. This is performed using an incompressible and inviscid model which gives a retroaction effect on the structure proportionally to its velocity. The two first parts of this text are devoted to a formulation of this three effects which are necessary in the dynamic modeling of a flexible (or not) structure immersed in a flow (air or water for instance). Examples in civil engineering and aerodynamics are given in order to

P. Destuynder (✉)
LaSIE, Univ-La Rochelle, La Rochelle, France
e-mail: philippe.destuynder@univ-lr.fr

C. Fabre
LMO-UMR 8628, CNRS Univ-Saclay, Univ- Paris-Sud, Orsay, France
e-mail: caroline.fabre@u-psud.fr

© The Author(s), under exclusive license to Springer Nature Switzerland AG 2021
D. Greiner et al. (eds.), *Numerical Simulation in Physics and Engineering: Trends and Applications*, SEMA SIMAI Springer Series 24,
https://doi.org/10.1007/978-3-030-62543-6_1

illustrate the theoretical formulation. Few control aspects in a dynamic behavior of the coupled fluid-structure modeling are also discussed in a section of this text.

Keywords Aeroelasticity · Flutter · Limit cycle of oscillation · Control of vibrations · Instabilities

1 Introduction

The aeroelasticity science is born with the first aircraft built by the Wright brothers in 1902 and also from the very interesting study performed by Otto Lilienthal since 1890 who unfortunately died in a crash of his wind glider. The work of Gustave Eiffel should also be mentioned even if he mainly focused his contribution to aerostatic. The static stability of aircrafts was a corner stone on which a lot of energy was spent by engineers, but the dynamic stability was only discussed after the second world war. Furthermore, the apparition of electronic computers enables one to develop new algorithms opening a new area in aeroelasticity. The physical understanding of the phenomena and their validations from the experiment are necessary. One could say that this science is relevant from scientific empiricism and is strongly based on the reasoning using mathematical models. Let us mention several references which have been helpful to us: A.V. Balakrishnan [2], R.L. Bisplinghoff, H. Ashley and H. Halfman [9], and R. H. Scanlan [29], H.C. Curtiss Jr., R.H. Scanlan, F. Sisto, E.H. Dowell, [13], J. R. Wright, and J.E. Cooper [32], M.E. Hoque [21] and Y.C. Fung [18]. Concerning general articles on aeroelasticity one can consult A.R. Collar [12], I.E. Garrick, and W.H. Reed [19]. Few extensions of this text can be found in Ph. Destuynder [16].

We start this course with the example of the Tacoma Narrows bridge. It is the opportunity to give a brief survey of the classical phenomena known in aeroelasticity. Sections 2.6–2.9 are devoted to the most recent analysis of the Tacoma bridge collapse which is now known as a stall flutter phenomenon. An other example implying a test model in a wind tunnel for which we have precise measurements, shows the similarity with the bridge. It is given in Sect. 8. Several control aspects are also discussed along the text but mainly in Sect. 5 in which we define most classical approaches. In case of an instability, an important problem for the engineers is to decide if there is or not a limit cycle of oscillations. If there is one, it is necessary to characterize its size. This enables one to decide if a fatigue phenomenon can occur or not. We explain how to characterize and to compute such limit cycles of oscillations (if there is one) in Sect. 6. But first of all, we give in the following section, a discussion of this Tacoma Narrows bridge failure which is sometimes described as the birth of modern quasi-static aeroelasticity.

2 The Collapse of Tacoma Narrows Bridge

In this section, we use Den Hartog's theory [14] (which describe the preliminary movement of the bridge before the collapse). Then, the explanation suggested by Y. Billah and R. Scanlan which led to the final torsion movement is given [8]. This is also the opportunity for us, to give a summary of phenomena involved in aeroelasticity problems (Fig. 1).

2.1 What Has Been Observed

The story happened on November 11th 1940 at Tacoma Narrows. A nice suspended new bridge collapsed down after 40 min of oscillations. There was a wind which velocity was approximately 25 m/s and the period of oscillations was about 5 s. The width of the cross section of the bridge was about 10 m (see Fig. 2); hence the reduce frequency was approximately:

$$f_r = \frac{10 \text{ m}}{(25 \text{ m/s}) \times 5 \text{ s}} = 0.08 << 1, \tag{1}$$

which enables to ensure that quasi-static aeroelasticity can be apply.

The Reynolds number representing a ratio between the energy transferred and the diffused one is given by:

$$R_e = \frac{UL}{\nu}, \tag{2}$$

Fig. 1 Two professors who contribute to the Aeroelasticity science: Y. Fung and R. Scanlan

Fig. 2 The Tacoma bridge which collapsed down on the 11-7-1940 four months after its inauguration

where U is the average velocity of the flow, L the length of the structure swept by the flow and v the kinematical viscosity. Obviously it is only an approximation of physical phenomena, but it enables one to classify the type of flow. In a standard way one refers to Fig. 3 for the different types of flow with respect to the value of the Reynolds number. In the case of Tacoma Narrows bridge the Reynolds number is approximately 10^7. The period of oscillations observed from a movie taken by a man who was in a car stopped on the bridge and who escape from the accident, was about 5 s. No body died excepted the dog of the man which refused to get out of the car.

2.2 The Strouhal Instability

When the Reynolds number is small enough (less than 2000) some particular instabilities can appear in the boundary layer and vortices are shedded into the wake. They are called Strouhal instabilities and have been widely studied by Th. von Karman. The characteristic number used is the so-called Strouhal number and

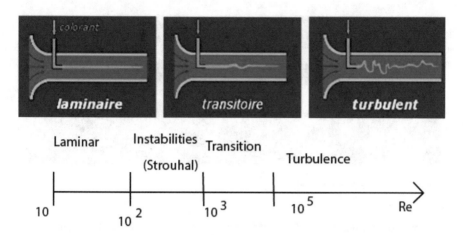

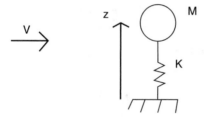

Fig. 3 Classification of flows versus the Reynolds number

Fig. 4 The simple model discussed for the Strouhal effect

is commonly defined by:

$$S = \frac{D}{U} f_s,$$ (3)

where U is the velocity of the flow, f_s is the frequency of the vortices shedding and D the dimension of the obstacle transversally to the flow direction. In the case of a cylinder—which main axis is transversal to the flow velocity—this Strouhal number is measured experimentally and is equal to .2 for this shape (a cylinder). Let us explain how this phenomenon can induce instabilities which could be assimilated to a resonance. This justifies why this terminology is correct.

Let us consider a cylinder of mass M as shown on Fig. 4.

The only movement authorized is the vertical displacement denoted by z. Because of the spring which stiffness is K, the equation of the movement is:

$$M\ddot{z} + Kz = f(t), \ z(0) = 0, \ \dot{z}(0) = 0,$$ (4)

where $f(t)$ is the reaction force applied to the cylinder and is due to the separation of the vortices from the boundary layers around the cylinder (see Fig. 5). The function

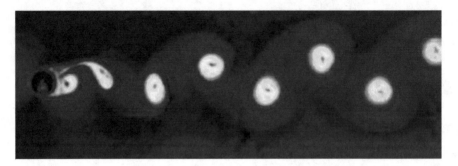

Fig. 5 The von Karman instabilities

f is periodic with a period $T_s = 1/f_s$ as far as there is a Strouhal instability of the boundary layers and one can consider for instance that:

$$f(t) = F \sin(2\pi f_s t), \tag{5}$$

Setting $\omega = \sqrt{\dfrac{K}{M}}$, the solution is:

$$z(t) = \frac{F}{M\omega} \int_0^t \sin(2\pi f_s s) \sin(\omega(t - s)) ds$$

$$= \frac{F}{M\omega((2\pi f_s)^2 - \omega^2)} [\omega \sin(2\pi f_s t) - (2\pi f_s) \sin(\omega t)]. \tag{6}$$

When $\omega \simeq 2\pi f_s$, one has:

$$z(t) \simeq \frac{F}{2M\omega^2} [\sin(\omega t) - t\omega \cos(\omega t)]. \tag{7}$$

The curves representing the solution z are plotted on Fig. 6 for different values of ω close to $2\pi f_s$. The frequency of vibration which was measured from the movie taken by the owner of the car left on the bridge, was about .2 Hz and the Strouhal frequency of a H-shaped cross section as the one of Tacoma bridge (see Fig. 9), is about $.1 \times \frac{D}{U} = .08$ Hz (the coefficient .1 is the Strouhal number and was measured in several wind tunnels) and $D \simeq 2$ m is the height of the girders. Therefore, there is a large difference between this two values. Anyway the forces implied in a Strouhal instability are small and can be cancelled by a small damping of the mechanical system. In addition -and mainly- the phenomenon occurs for very low Reynolds numbers (less than 10^4) and in the Tacoma Narrows bridge case the Reynolds number was about $\dfrac{U \times L}{\nu} \simeq 10^7$. This is why this hypothesis has not been kept, even if it was strongly supported by Th. von Karman. In order to try to converge to an agreement between all the members of the aeroelasticity community,

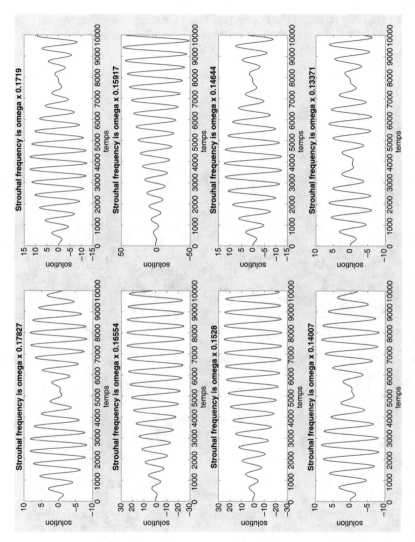

Fig. 6 Solutions of Eq. (4) for different values of f_s. In this example the resonance occurs for $f_s = \frac{\omega}{2\pi} \simeq 0.15917\,\text{Hz}$

one could say that *may be the Strouhal instability can be at the very origin of the phenomenon observed when the wind velocity was about* 1 m/s.

2.3   The Resonance of a Bridge

In 1850, in the city of Angers in France (see Fig. 7), a bridge over the river *La Loire* collapsed due the passage of soldiers walking at pace. Unfortunately the frequency

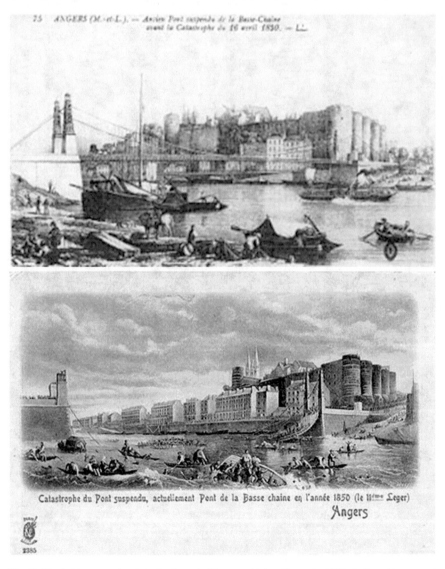

Fig. 7 The bridge over the river *La Loire* in France which collapse in 1850: before (above) and after (below)

of the pace was one of the eigenfrequency of the bridge which enter in resonance and collapsed. Since that date, it is forbidden to walk at pace on any bridge in the world. In the case of Tacoma Narrows bridge the frequencies have been computed very accurately using finite element codes by different research teams in the world. The one observed in the final movement before the destruction seems to be different and the spectrum of the turbulent flow observed in this district (near Tacoma in the south part of Seattle which is in the state of Washington) is not very stable in time as usual in such a phenomenon. Therefore the resonance, which is a linearly increasing instability as shown on Fig. 6, can not be triggered by this turbulent flow. Hence, this explanation is not strongly propped and was given up by most of the engineers (but not by several journalists who found this simple—but false—theory very attractive).

2.4 The Buffting Phenomenon

During the second world war, the military aircrafts met several strange phenomena. One was the buffting (mainly observed on the so called Wildcat aircrafts). It is due to vortices shedding from the main wings on the rear wings which induces a periodic excitation. The movement of the rear wing can be modeled in a first step, by a Hill's equation (see for instance the book by M. Roseau [28]) as follows:

$$\ddot{\alpha} + \omega^2(1 + a(t))\alpha = a_0, \ \alpha(0) = 0, \ \dot{\alpha}(0) = 0, \tag{8}$$

where α is the amplitude of an eigenmode of the rear wing. One can prove that if a is a periodic function which period is close to $\frac{2\pi}{\omega}$ (and not necessarily exactly equal), then the solution α increase exponentially with respect to the time. In fact non linear terms can reduce the magnitude of α. This is the buffting phenomenon. In the case of the Tacoma Narrows bridge, it would have been necessary that the energy spectrum of the wind turbulence contains meaningful terms at the frequency corresponding the movement observed on the movie. This is a quite low frequency which should correspond to large vortices in the flow. This phenomenon has not been registered by the weather forecasting station in the Tacoma valley and therefore it seems difficult to imply this buffting instability in the collapse of the bridge.

2.5 Flutter Induced by a Coupling Between Two Eigenmodes

The classical flutter mechanism is more complex and has been detected since the beginning of aeronautic but its understanding is more recent. One can refer to an history of flutter by I.E. Garrick and W.H. Reed [19]. It is due to a coupling between two eigenmodes of a flexible structure. In fact, this coupling appears when a double eigenfrequency situation occurs. It depends obviously on the conception of the structure itself, but also on the velocity of the flow and on its evolution with respect to the position of the structure with respect to the main flow direction. When a double eigenfrequency appears, it is possible that an imaginary eigenvalue is developed and leads to an exponential instability. In this case, the eigensubspace is two dimensional. One eigenmode capture energy from the flow and transfers it to the other one which stores it creating the instability. Let us give a simple and classical example.

2.5.1 A Short Description of Flutter Phenomenon

Let us consider a two dimensional airfoil as shown on Fig. 8. It is fixed by two springs: one is a traction spring and the second one is a torsion spring. The stiffness are respectively k and c. Two movements are possible. One is the heaving, denoted by z and the other one is the pitching denoted by α. The linearized equations of the model around $z = \dot{z} = \alpha = \dot{\alpha} = 0$, are the following one and can be easily derived from the expression of the Lagrangian. Let us introduce few notations.

- M is the mass of the airfoil;
- J_0 its inertia around point O, J_G is the inertia around point G, the center of mass;
- V is the flow velocity far away from the structure;
- a is the algebraic distance between O and G;
- c_x, c_z are respectively the drag and the lift coefficients. The corresponding aerodynamical forces are $\frac{\rho S V^2}{2} c_x$ and $\frac{\rho S V^2}{2} c_z$, S is a cross section used as a reference surface;

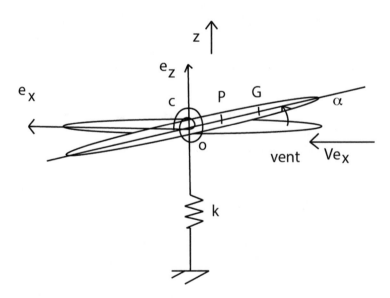

Fig. 8 The system considered

- c_m is the pitching coefficient and the pitching moment is $\frac{\rho S L V^2}{2} c_m$, L is a characteristic length.

All the aerodynamic coefficients are assumed to be known in this presentation. They can be obtained from a wind tunnel test or from a computer code. In fact, there are six coefficients (the three other are c_y for the sheering, c_r for the rolling and c_l for the yawing). All these coefficients are functions of the angles characterizing the position of the structure: α for the pitch, β for the yaw and γ for the roll. But we essentially use c_x, c_z and c_m in this text.

The lagrangian of the mechanical system is with (the springs are assumed to be at rest at the origin):

$$\begin{cases} L(\dot{z}, z, \dot{\alpha}, \alpha) = E_c - E_p, \\[2mm] Ec = \tfrac{1}{2}(M[\dot{z}^2 + 2a\dot{\alpha}\dot{z}\cos(\alpha)] + J_0\dot{\alpha}^2) \\[2mm] E_p = \tfrac{1}{2}(kz^2 + c\alpha^2) + Mg(z + a\alpha). \end{cases}$$

Therefore the dynamical model consists in finding z and α such that:

$$\begin{cases} \dfrac{d}{dt}(\dfrac{\partial L}{\partial \dot{z}}) - \dfrac{\partial L}{\partial z} = \dfrac{\rho S V^2}{2} c_z(\alpha) - Mg, \\[4mm] \dfrac{d}{dt}(\dfrac{\partial L}{\partial \dot{\alpha}}) - \dfrac{\partial L}{\partial \alpha} = \dfrac{\rho S L V^2}{2} c_m(\alpha) - aMg. \end{cases} \tag{9}$$

or else from a simple computation:

$$\begin{cases} M\ddot{z} + a\cos(\alpha)M\ddot{\alpha} - a\dot{\alpha}^2\sin(\alpha)M + kz = \dfrac{\rho S V^2}{2} c_z(\alpha) - Mg, \\[4mm] J_0\ddot{\alpha} + a\cos(\alpha)M\ddot{z} + c\alpha = \dfrac{\rho S L V^2}{2} c_m(\alpha) - aMg. \end{cases} \tag{10}$$

And after a linearization near $\alpha = 0$:

$$\begin{cases} M\ddot{z} + aM\ddot{\alpha} + kz = \dfrac{\rho S V^2}{2}[c_z(0) + \dfrac{\partial c_z}{\partial \alpha}(0)\alpha] - Mg, \\[4mm] J_0\ddot{\alpha} + aM\ddot{z} + c\alpha = \dfrac{\rho S L V^2}{2}[c_m(0) + \dfrac{\partial c_m}{\partial \alpha}(0)\alpha] - aMg. \end{cases} \tag{11}$$

In a matrix form one obtains (with $R = \dfrac{\rho S V^2}{2}$ and $Q = \dfrac{\rho S L V^2}{2} = RL$):

$$
\overbrace{\begin{pmatrix} M & aM \\ aM & J_0 \end{pmatrix}}^{M \text{ inertia matrix}} \begin{pmatrix} \ddot{z} \\ \ddot{\alpha} \end{pmatrix} + \overbrace{\begin{pmatrix} k & -R\dfrac{\partial c_z}{\partial \alpha}(0) \\ 0 & c - Q\dfrac{\partial c_m}{\partial \alpha}(0) \end{pmatrix}}^{K \text{ stiffness matrix}} \begin{pmatrix} z \\ \alpha \end{pmatrix} = \overbrace{\begin{pmatrix} Rc_z(0) - Mg \\ Qc_m(0) - Mga \end{pmatrix}}^{F \text{ aero-forces}} \tag{12}
$$

Remark 1 If the pitching coefficient c_m is positive the resultant aerodynamic force is forward the center of rotation ($a > 0$) and backward ($a < 0$) if it is negative. Furthermore, the torsor of aerodynamic forces is assumed to equilibrate the weight in a standard flight. Hence the right-hand side of the previous equation is zero. □

2.5.2 Linear Stability Analysis

The stability of the model rests on the imaginary part of the eigenvalues λ solution of the equation:

$$
\det(\lambda M - K) = 0. \tag{13}
$$

For sake of simplicity in the notations, we set::

$$
\omega_1 = \sqrt{\dfrac{k}{M}}, \quad \omega_2 = \sqrt{\dfrac{c}{J_G}}, \quad E = \dfrac{\rho S L}{2}\dfrac{\partial c_m}{\partial \alpha}(0), \quad D = \dfrac{\rho S}{2}\dfrac{\partial c_z}{\partial \alpha}(0), \quad \xi = \dfrac{J_0}{J_G} \geq 1. \tag{14}
$$

The previous equation with respect to λ is:

$$
\lambda^2 - \lambda\left(\omega_1^2 \xi + \omega_2^2 + \dfrac{aD - E}{J_G}V^2\right) + \omega_1^2\left(\omega_2^2 - \dfrac{EV^2}{J_G}\right) = 0 \tag{15}
$$

The pulses are $\mu = \pm\sqrt{\lambda}$. And the elementary solutions of the linearized model (12) are exponential as $e^{i\mu t}$. Hence the stability rests on the imaginary part of μ. The system is unstable as soon as there is a root in μ with a negative imaginary part.

- If $\lambda > 0 \Longrightarrow \mu \in \mathbb{R} \Longrightarrow$ stable (both static and dynamic),
- If $\lambda < 0 \Longrightarrow \mu \in \mathbb{C} \Longrightarrow$ unstable (static),
- If $Im(\lambda) \neq 0 \Longrightarrow \mu \in \mathbb{C} \Longrightarrow$ unstable in dynamic $\Longrightarrow$ *flutter*.

2.5.3 Discussion on the Apparition of the Flutter (Roots in λ Are Real or Complex Conjugate)

- If $V = 0 \Rightarrow \text{sign}(\Delta) = \text{sign}(\omega_1^2(\xi - 1) + (\omega_1 - \omega_2)^2) \Rightarrow \lambda \in \mathbb{R}^{+*} \Rightarrow$ stable

- If $E > 0$ and $V > V_c = \omega_2\sqrt{\dfrac{JG}{E}} = \sqrt{\dfrac{c}{E}} \Rightarrow \lambda_1\lambda_2 < 0 \Rightarrow$ unstable

- If $V < V_c$, let us set $\Delta = AV^4 + 2BV^2 + C, (C > 0)$

$*$ If $\Delta > 0 \Rightarrow$ two real roots with the same sign $\Rightarrow$ stable

A particular important case corresponds to $aD = E$:

1. $*$ If $aD = E \Rightarrow \Delta = 2BV^2 + C$; therefore if $E > 0 \Rightarrow B > 0 \Rightarrow$ stable

2. $*$ If $aD = E$ and $E < 0$ stable if $V < V_{f_0} = \dfrac{1}{2\omega_1}\sqrt{\dfrac{CJG}{-E}}$

3. **General case.** Let us set:

$$\begin{cases} A = \dfrac{(aD-E)^2}{J_G^2}, \quad B = 2\dfrac{\omega_1^2 E}{J_G} + (\omega_1^2\xi + \omega_2^2)\dfrac{aD-E}{J_G}, \\[2mm] C = (\omega_1^2\xi + \omega_2^2)^2 - 4\omega_1^2\omega_2^2. \end{cases}$$

We introduce:

$$\delta' = B^2 - AC = \dfrac{4\omega_1^2}{J_G^2}[E^2(\omega_1^2(1 - \xi) + EaD(\omega_1^2\xi - \omega_2^2) + a^2D^2\omega_2^2]$$

which implies that (Fig. 9):

$*$ if $\delta' < 0 \Rightarrow$ stable,

$*$ if $\delta' > 0 \Rightarrow$ *flutter* if $V_{f_e} < V < V_{f_s}$; V_{f_e}, V_{f_s} roots of Δ.

2.5.4 Graphic Interpretation of the Flutter

The eigenfrequencies in the complex plane versus the velocity of the flow have been plotted on Fig. 10. On the left top of Fig. 10, we have plotted the discriminant δ' and on the right top we have drawn Δ. One can see a flutter area localized between two velocities V_{f_1} et V_{f_2} which are roots of $\Delta = 0$.

Bottom left of this figure, one can see the evolution of the real part of λ and bottom right the imaginary part of λ has been plotted.

The *flutter* corresponds to a crossing of two eigenfrequencies (one for the bending and the other for the torsion). But when the velocity is higher, the flutter disappears. Hence one can speak of a flutter range corresponding to the interval

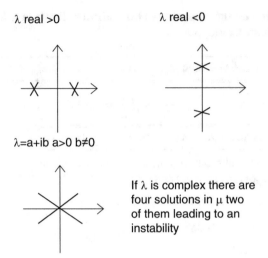

Fig. 9 Roots in $\mu = \pm\sqrt{\lambda}$

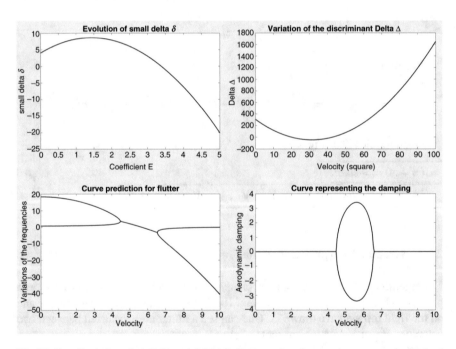

Fig. 10 Top: Evolution of Δ (left) and δ (right). Bottom: eigenfrequencies versus velocity (real part left, imaginary part right)

between the two velocities for which Δ. The numerical simulations has been performed with $a < 0$ (G behind O) in order to avoid some other instabilities.

2.5.5 Discussion of the Flutter Phenomenon for the Tacoma Narrows Bridge

The basic point in the detection of flutter is the fact that two eigenfrequencies should be very close and get narrow and narrow due to the effect of the aerodynamic forces (aerodynamic stiffness). In the case of Tacoma Narrows bridge, a very precise finite element analysis of the spectrum performed by different research team across the world (in particular at the National Aerospace Laboratory in Tokyo) have eliminated this possibility. Even if, many teaching books suggest that the flutter could be responsible of the collapse of the bridge, this is not correctly propped. Furthermore, the movie suggests without doubt a transition from galloping (bending of the bridge) to a torsion, but certainly not a coupling between these two movements. Hence, some other explanations were necessary. In fact, it is the stall flutter which remains the only trustworthy phenomenon as responsible of the collapse.

2.6 The Galloping of the Bridge and the Stall Flutter Phenomenon

In this subsection, we develop the analysis based on the apparent wind velocity which is known today as the right explanation of Tacoma bridge breakdown. Let us give a short description of this theory. The first step is the initiation of the movement which commonly is due to the effect of the apparent wind velocity on the bending of the bridge. The basic point for this phenomenon was based on the famous criterion suggested by J. Den Hartog (Fig. 11). The corner stone of Den Hartog theory, is to define a new Eiffel frame linked to the direction of the apparent wind. Therefore the apparent position of the structure with respect to the wind, now depends on the velocity of the structure at the point where the aerodynamic forces are applied (if they exist). In case of a translation there is no moment implied, hence these forces are reduced to a resultant and because all the point of the structure have the same velocity, there is no ambiguity to apply this resultant force at the center of mass of the structure.

Let us use the two systems of axis represented on Fig. 12. The first one $(O; e_1, e_2, e_3)$ is connected to the absolute wind direction ($v = Ve_1$) and the second

Fig. 11 Find above a picture of Professor Den Hartog. J.P. Den Hartog was born in 1901 in Ambarova, the Dutch East Indies. He enrolled at Delft University of Technology in 1919 and received his MSc degree in electrical engineering in 1924. He emigrated to the United States in 1924. From 1932 to 1945 he taught at Harvard University. From 1945 to 1967 he taught dynamics and strength of materials at MIT in the Department of Mechanical Engineering. In 1962 he became jointly appointed as a professor in the Department of Naval Architecture and Marine Engineering. Jacob Pieter Den Hartog died at the age of 87 on March 17, 1989

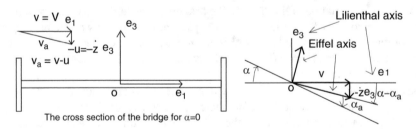

Fig. 12 The cross section of a H-shape bridge and the two systems of axis used for computing the Den Hartog criterion

one is connected to the apparent wind, is denoted by $(O; e_1^a, e_2^a, e_3^a)$. The apparent wind velocity expressed in the first frame is:

$$v^a = V e_1 - \dot{z} e_3, \tag{16}$$

the apparent angle of attack α^a (angle between the apparent wind and e_1) is therefore:

$$\alpha^a = \alpha - \arctan(\frac{\dot{z}}{V}), \tag{17}$$

and the aerodynamic forces expressed in the Eiffel frame linked to the apparent wind are:

$$F_z(\dot{z}) = \frac{\rho S ||v^a||^2}{2} \left[c_x(\alpha^a)e_1^a + c_z(\alpha^a)e_3^a \right].e_3, \tag{18}$$

where the axis linked to the apparent wind are given by:

$$\begin{cases} e_1^a = \cos(\alpha - \alpha^a)e_1 - \sin(\alpha - \alpha^a)e_3, \\ \\ e_3^a = \sin(\alpha - \alpha^a)e_1 + \cos(\alpha - \alpha^a)e_3. \end{cases} \tag{19}$$

Hence, the expression of F_z becomes:

$$F_z(\dot{z}) = \frac{\rho S ||v^a||^2}{2} \left[-\sin(\alpha - \alpha^a)c_x(\alpha^a) + \cos(\alpha - \alpha^a)c_z(\alpha^a) \right]. \tag{20}$$

The non linear model is (the pitching angle α appears as a parameter and is fixed but α_a depends on $\dot{z}$):

$$M\ddot{z} + Kz = F_z(\dot{z}), \quad z(0) = 0, \quad \dot{z}(0) = 0. \tag{21}$$

One can build the linearized expression of this model.

Let us first observe that one has near $\dot{z} = 0$ and $\alpha = \alpha_0$:

- $F_z(\dot{z}) = F_z(0) + \dot{z}\dfrac{\partial F_z}{\partial \dot{z}}(0) + \ldots,$

- $F_z(0) = \dfrac{\rho S V^2}{2} c_z(\alpha_0),$

- $\dfrac{\partial \alpha^a}{\partial \dot{z}}(0) = -\dfrac{1}{V},$

- $\dfrac{\partial ||v^a||^2}{\partial \dot{z}}(0) = 0,$

- $\dfrac{\partial F_z}{\partial \dot{z}}(0) = -\dfrac{\rho S V}{2} \left[c_x(\alpha_0) + \dfrac{\partial c_z(\alpha_0)}{\partial \alpha} \right].$

Therefore the linearized model around $\dot{z} = 0$ and $\alpha = \alpha_0$ (and without initial perturbation for sake of brevity) is (M is the mass matrix, D is the so-called aerodynamic damping and K is the stiffness matrix in the direction z):

$$\begin{cases} \bullet \ M\ddot{z} + D\dot{z} + Kz = F_z(0), \ z(0) = 0, \ \dot{z}(0) = 0, \\ \\ \bullet \ D = \dfrac{\rho S V}{2} \left[c_x(\alpha_0) + \dfrac{\partial c_z(\alpha_0)}{\partial \alpha} \right]. \end{cases} \quad (22)$$

The stability of the solution of the previous equation rests on the sign of the damping coefficient D which is a function of the angle of attack α_0 and not on the velocity of the flow V. Let us discuss what can happen.

1. If $D > 0$ the system is damped through an aeroelastic phenomenon,
2. If $D = 0$ there is no damping and the system is conservative (excepted if there is a structural damping),
3. If $D < 0$ the system is unstable (we do not take into account a possible structural damping). In this case, the solution is exponentially increasing with the time.

In all cases, the drag coefficient c_x is positive. Concerning the lift coefficient c_z the situation is more complicated. Even if it is positive, a stall phenomenon can occur (corresponding to a local maxima ($\dfrac{\partial c_z}{\partial \alpha} = 0$)). If $\dfrac{\partial c_z}{\partial \alpha}$ is sufficiently negative (which can appear after or before the stall), a negative damping appears and therefore there is an instability of the system.

In the case of the Tacoma Narrows bridge, the curves of the aerodynamic coefficient are those plotted on Fig. 13.

The decrease of the lift coefficient for α small is surprising but it leads to a negative value of D. It is due to the following phenomenon. At each extremity of the H (see Fig. 14) there are vortices which are developed symmetrically inside the two cavities (4 vortices!). But for even a small inclination of the H with respect to the flow direction, one of the bottom vortex is blow out from the cavity and therefore the pressure decreases. This explains the strange behavior of the lift as shown on Fig. 13. It is commonly accepted by most prominent engineers in civil engineering that this stall flutter instability is at the origin of the bridge movement, but not responsible of the galop observed at the end, until the breakdown. Another phenomenon appeared after that one. It is described in the next subsection; it is the most important one and is known as responsible of the collapse of the bridge.

2.7 Letter from W.P. Rodden to R. Scanlan About Tacoma

See Fig. 15.

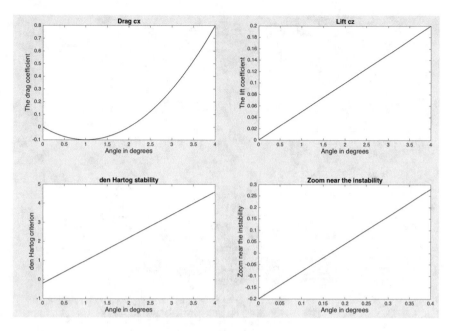

Fig. 13 Aerodynamic coefficients of a H shape cross section. First line c_x, c_z and second line D and a zoom near $D = 0$ (the instability occurs for $D < 0$)

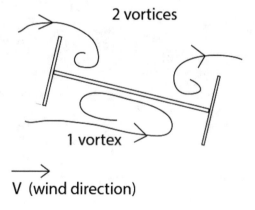

Fig. 14 The vortices developed in the cavities of the H shape

2.8 Answer from R. Scanlan to W.P. Rodden About Tacoma

See Fig. 16.

WILLIAM P. RODDEN, PH.D., INC.
CONSULTING ENGINEER
255 STARLIGHT CREST DRIVE
LA CAÑADA FLINTRIDGE, CALIFORNIA 91011

5 March 1991

Dr Robert H. Scanlan
Department of Civil Engineering
202 Latrobe Hall
The Johns Hopkins University
Baltimore, MD 21218-2699

Dear Bob:

I am somewhat overdue in sending your books for your autograph as we discussed last November at SWRI at the AFDC (Flutter Club) cocktail party. I enjoyed very much meeting you after all these years. Again, I thank you (after 35 years!) for the inspiration your first book gave for my earliest work on flutter analysis by influence coefficient methods -- it suggested the idea of aerodynamic influence coefficients to me in 1956 that eventually worked its way into the aeroelastic features we have in NASTRAN today.

I also enjoyed hearing your story about your efforts to educate Lazer and McKenna about the aerodynamic origin of bridge flutter. They are rather hopeless cases. I have also enclosed some results of an implementation of your bridge aerodynamics into MSC/NASTRAN. It seems strange that all 8 of the vibration modes would go unstable at some speed. Would you suspect that some (or all) of the roots are wrong, or would you expect a whole series of single degree of freedom instabilities?

Your news about Bob Rosenbaum's death was sad. He was a great pioneer and has left a great heritage. Your book together is still a classic. Some of my correspondence with him is inside the front cover. Again, I enjoyed meeting you and will be grateful for your autographs. Best regards.

Sincerely yours,

Bill

William P. Rodden

Fig. 15 Letter from W.P. Rodden to R. Scanlan about Tacoma bridge

2.9 The Stall Flutter Phenomenon in the Torsion of the Bridge

Let us consider in this subsection that the movement of the cross section of the bridge is a rotation around a fixed point O. The angle of rotation is denoted by α (see Fig. 17).

THE JOHNS HOPKINS UNIVERSITY — G. W. C. WHITING SCHOOL OF ENGINEERING

Department of Civil Engineering, Baltimore, Maryland 21218-2699 (301) 338-8680

March 13, 1991

Dr. William P. Rodden
255 Starlight Crest Dr.
La Cañada Flintridge, CA 91011

Dear Bill:

Returned herewith are your copies of the two books you asked me to autograph. Done, and thank you for asking. The now somewhat historic letters from Bob Rosenbaum are approaching collector's item status.

The Japanese bridge flutter paper is of interest, particularly in incorporating experimental flutter derivatives into MSC/NASTRAN. I have not followed out all the details implied in what the Japanese did. I can comment, however, that the heart of the matter re Tacoma Narrows 1940 is the essential s.d.o.f. instability due to negative damping in twist (cf. coefficient A_2^*), particularly as manifested in the fundamental torsion mode at $0.2\,Hz$. Enclosed is a very recent paper I did with one of my Princeton students on the O.T.N. for the American Journal of Physics— 50 years after the event. Therein we offer what I consider pretty convincing evidence that it was essentially this single antisymmetric torsion mode that drove that span to destruction.

When the full set of flutter derivatives is used in an analysis, I think some aerodynamic coupling-in of other modes may be demonstrated. In any event, the highly unstable twist tendency of the O.T.N. deck cross section will dominate. We found flutter speeds of 17 to 30 mph, depending on damping, in our single-mode analysis, as our AJP paper shows. In this connection the Japanese result— a critical speed of 12 m/s (27 mph) for $g = 0.01$ (our $\zeta = 0.005$) is in the ball park but slightly higher. The Japanese may have been inexact as to the modes of the O.T.N. bridge: for us, the lowest critical mode was the torsion at about 12 cpm (0.2Hz) or about

$$\sqrt{50} \times 0.2 Hz = 1.4 Hz$$

in the 1:50 scale model tested by Farquharson years ago (see our Fig.3).

Thanks for your letter. It was nice seeing you in Texas. Best regards.

Yours sincerely,

Robert H. Scanlan

Fig. 16 Answer of R. Scanlan to W.P. Rodden still about Tacoma bridge

The elements of reduction of the aerodynamic forces applied to the cross section of the bridge expressed at point O which is a fixed point, are the resultant:

$$R = \frac{\rho S V^2}{2}\big[c_{0x}(\alpha)e_1 + c_{z0}(\alpha)e_3\big],$$

and the pitching moment:

$$M = \frac{\rho S L V^2}{2}c_{0m}(\alpha)e_2.$$

Fig. 17 Torsion movement for the bridge

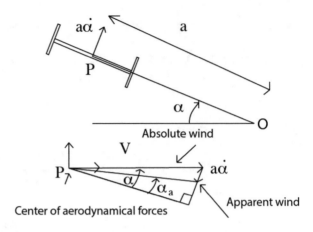

There exists a point P localized on the axis containing the point O and the deck of the bridge (see Fig. 17) such that the moment is zero. This point is characterized by the following relation where a is the distance from P to O:

$$c_{Pm}(\alpha) = c_{0m} - a\big[c_{0x}(\alpha)\sin(\alpha) + c_{z0}(\alpha)\cos(\alpha)\big] = 0, \tag{23}$$

or else:

$$a = \frac{c_{0m}}{\big[c_{0x}(\alpha)\sin(\alpha) + c_{z0}(\alpha)\cos(\alpha)\big]}. \tag{24}$$

In many cases, the distance a is constant for α small enough as far as all the aerodynamic coefficients vary linearly with respect to the angle of attack α. In the case of Tacoma Narrows bridge, this is true excepted in a small neighborood of the origin (see Sect. 2.6). The apparent wind at this point P is:

$$V^a(P) = V e_1 - a\dot{\alpha}\big[\cos(\alpha)e_3 + \sin(\alpha)e_1\big]. \tag{25}$$

and therefore the modulus and the apparent angle of attack at P are:

$$\begin{cases} \|V^a(P)\|^2 = V^2\Big(1 - \dfrac{2a\dot{\alpha}}{V}\sin(\alpha) + \big(\dfrac{a\dot{\alpha}}{V}\big)^2\cos(\alpha)^2\Big), \\[4mm] \alpha^a = \arctan\Big(\tan(\alpha) - \dfrac{a\dot{\alpha}}{V\cos(\alpha)}\Big). \end{cases} \tag{26}$$

The linearization around $\alpha = \alpha_0$ leads to:

$$\begin{cases} \alpha^a = \alpha_0 - \dfrac{a\dot{\alpha}\cos(\alpha_0)}{V} + \dots \\[4mm] ||v^a||^2 = V^2 - 2aV\dot{\alpha}\sin(\alpha_0) + \dots \end{cases} \tag{27}$$

The moment of the aerodynamic forces at O is therefore:

$$M_0(\alpha, \dot{\alpha}) = \frac{\rho SL||v^a||^2}{2} c_{0m}(\alpha^a)$$

$$= \frac{\rho SLV^2}{2}\Big(c_{0m}(\alpha_0) - \frac{a\dot{\alpha}}{V}[2\sin(\alpha_0)c_{0m}(\alpha_0) + \cos(\alpha_0)\frac{\partial c_{0m}}{\partial \alpha}(\alpha_0)] + \dots\Big). \tag{28}$$

Finally, if C is the stiffness of the torsion spring maintaining the system (see Fig. 17) and J_0 the inertia moment of the system at O, the linearized equation of the movement around α_0 which is defined by (here C is the stiffness of the torsion spring):

$$C\alpha_0 = \frac{\rho SLV^2}{2} c_{0m}(\alpha_0), \tag{29}$$

is:

$$J_O\ddot{\alpha} + \frac{\rho SLV}{2}[2\sin(\alpha_0)c_{0m} + \cos(\alpha_0)\frac{\partial c_{0m}}{\partial \alpha}(\alpha_0)]\dot{\alpha} + C(\alpha - \alpha_0) = 0. \tag{30}$$

We set:

$$D = \frac{\rho SLV}{2}[2\sin(\alpha_0)c_{0m}(\alpha_0) + \cos(\alpha_0)\frac{\partial c_{0m}}{\partial \alpha}(\alpha_0)]. \tag{31}$$

The stability of the solution rests on the sign of the damping coefficient—say D—which is only due to the aerodynamic in this case. One can state the Den Hartog criterion as follows.

Den Hartog Criterion

- **If $D < 0$ the system is unstable and the growth of the instability is exponential;**
- **if $D = 0$ the system is conservative (no damping);**
- **if $D > 0$ the system is aerodynamically damped.**

Remark 2 The coefficient D is proportional to the absolute velocity V. Hence for a small value, the structural damping can cancel an instability. □

Remark 3 Let us consider the ordinary differential equation:

$$2 \sin(\alpha) q(\alpha) + \cos(\alpha) \frac{\partial q}{\partial \alpha}(\alpha) = 0.$$

The solutions are:

$$q(\alpha) = \frac{q(\alpha_0)}{\cos(\alpha_0)^2} \cos(\alpha)^2,$$

where $q(\alpha_0)$ is an arbirary constant. The curve representing q have been plotted on Fig. 18 for various values of $A = \dfrac{q(\alpha_0)}{\cos(\alpha_0)^2}$. Let us consider a point Q on the curve representing c_{m0}. and corresponding to the angle of attack α_Q. One can choose A such that:

$$q(\alpha_Q) = c_{0m}(\alpha_Q). \tag{32}$$

This gives:

$$A_Q = \frac{c_{0m}(\alpha_Q)}{\cos(\alpha_Q)^2}. \tag{33}$$

Finally, let us plot the curves $\alpha \rightarrow q(\alpha)$ on the same graph as the one of c_{0m} (chosen artificially) for several values of α_Q.

□

In the case of Tacoma Narrows bridge, the instability in torsion has appeared for pitching angles approximately larger than 15°. The stall phenomenon (decrease of the pitching coefficient c_{0m} is due to the suppression of the large vortices in the lower cavity of bridge created by the deck and the girders. In fact a limit cycle of oscillations (positive damping for $\alpha \in [-15°, 15°]$ and negative damping for $|\alpha| > 15°$) has been observed inducing a fatigue phenomenon and the break down of the bridge (see Fig. 19). Let us underline that since this accident, it is forbidden to build bridges with full lateral girders (Fig. 20). Finally, let us underline that this analysis rests on the property that the point P (center of aerodynamic forces) does not moves too much. This is just an approximation and we discuss this important point in another section concerning a test that we performed in a wind tunnel for a military aircraft.

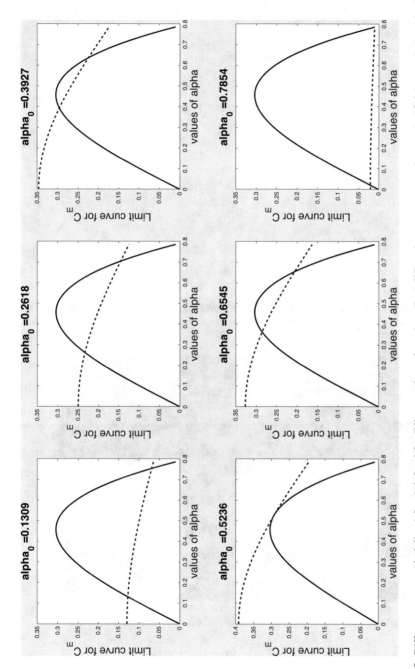

Fig. 18 Different curves q (dash lines) for A defined for different values of α_Q (in radian). When the curve representing c_{0m} (artificial example here) is above at the intersection point, the system is stable and unstable in the other case. The limit between the two possibilities occurs when the curves are tangent. From the practical point of view, the results given here are easy to use without computation but just with a graphic trick

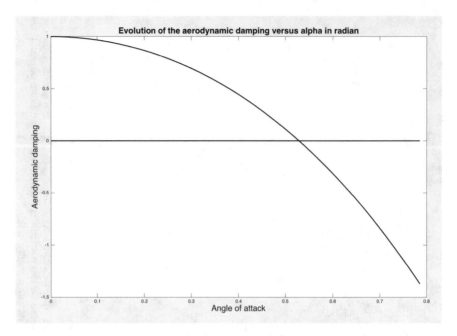

Fig. 19 The aerodynamic damping for the torsion movement with the coefficient c_{0m} used for the plotting of Fig. 18. The instability occurs when the damping becomes negative which corresponds to case of tangent curves on Fig. 18

Fig. 20 Torsion of the Tacoma bridge; picture colorized and taken from the true movie

3 Generalization of the Apparent Wind Method

In this section we generalize the method introduced in the previous section for simple examples. The basic idea is to split a complex structure into several pieces which are assumed to be separately rigid but articulated between them. Let us consider a rigid body with six degrees of freedom: three translations at a point O: $(d(O) = (d_1, d_2, d_3))$ and three rotations: $r = (r_1, r_2, r_3)$. The velocity of a point P of this body is:

$$\dot{d}(P) = \dot{d}(O) + \dot{r} \wedge OP. \tag{34}$$

The forces applied to this body is a torsor $\mathscr{T} = (F, M(O))$ where F is the resultant and $M(0)$ the moment at a point O. The moment at P is given by:

$$M(P) = M(O) + F \wedge OP. \tag{35}$$

The instantaneous power of these external forces applied by the flow on the body is (the dot is the scalar product in $\mathbb{R}^3$):

$$\mathscr{P} = F.\dot{d}(O) + M(0).\dot{r} = F.\dot{d}(P) + M(P).\dot{r} \tag{36}$$

The magnitude of the moment at point P is therefore: ($\langle ., ., \rangle$ is the mixed product in $\mathbb{R}^3$)

$$J(OP) = ||M(P)||^2 = ||M(O)||^2 + 2\langle M(O), F, OP \rangle + ||F \wedge OP||^2. \tag{37}$$

It is a bilinear, symmetrical and positive functional with respect to OP. Hence the minimum is obtained for a point P such that the gradient is vanishing.

Let us set:

$$F = \begin{pmatrix} F_1 \\ F_2 \\ F_3 \end{pmatrix}, \quad M(0) = \begin{pmatrix} M_1 \\ M_2 \\ M_3 \end{pmatrix}, \quad OP = X = \begin{pmatrix} x_1 \\ x_2 \\ x_3 \end{pmatrix}.$$

One has:

$$||F \wedge X||^2 = (AX.X)_3 \text{ where } A = \begin{pmatrix} F_3^2 + F_2^2 & -F_1 F_2 & -F_1 F_3 \\ -F_1 F_2 & F_3^2 + F_1^2 & -F_2 F_3 \\ -F_1 F_3 & -F_2 F_3 & F_1^2 + F_2^2 \end{pmatrix} = ||F||_2^2 I_3 - F\,{}^t F.$$

Let us introduce the notations:

$$B = F \wedge M(0) = \begin{pmatrix} F_2 M_3 - F_3 M_2 \\ F_3 M_1 - F_1 M_3 \\ F_1 M_2 - F_2 M_1 \end{pmatrix}.$$

The vector OP which minimizes the magnitude of $M(P)$ is solution of (F span the kernel of A and one has $B.F = 0$ and we prescribe $F.X = 0$ in order to have a unique solution):

$$AX = B \text{ with } F.X = 0.$$

In fact, for any $\varepsilon > 0$, solving the previous system is equivalent to solve:

$$\left(A + \varepsilon F^t F \right) X = B,$$

or else:

$$\left(I_3 + \frac{(\varepsilon - 1)}{\|F\|_2^2} F^t F \right) X = \frac{B}{\|F\|_2^2}.$$

Choosing for instance $\varepsilon = 1$, one obtains:

$$X = OP = \frac{B}{\|F\|_2^2} = \frac{F \wedge M(0)}{\|F\|_2^2}. \tag{38}$$

The minimum value of the moment at P is:

$$J(OP) = \|M(0)\|^2 - \langle F, M(0), OP \rangle. \tag{39}$$

The absolute wind velocity in the frame $(0; e_1, e_2, e_3)$ is denoted by $v = v_1 e_1 + v_2 e_2 + v_3 e_3$. But using Eiffel frame (e_1 is in the direction of the absolute wind), therefore one has $V e_1$. Once the point P has been characterized, we define the apparent wind for the body using the velocity at point P by;

$$v^a(P) = V e_1 - \dot{d}(P) = V e_1 - \dot{d}(0) - \dot{r} \wedge OP. \tag{40}$$

From this expression, we define the apparent angles of pitching, yawing and rolling as follows. Let first point out that we choose to define the three angles in the Eiffel frame. Hence even if they are equivalent to the Euler angles for instance, they are different and the change is classical (see Fig. 21).

As far in most applications, these angles are perturbations of a nominal angles, they can be close to zero (but not always). Hence the sign is important. This is why we prefer to use the vector product instead of the scalar product which would characterize the cosine of the apparent pitching angle α^a, (and similarly for the

Fig. 21 Euler angles ($\neq$ from those used in this text) (θ, φ, ψ)

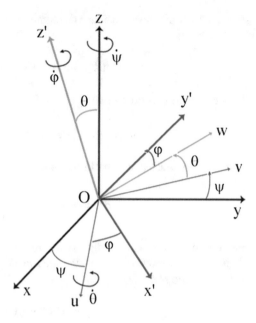

yawing and the rolling angles). First of all, the component of the apparent velocity in the plane (e_1, e_3) is used for defining the pitching angle. We set (projection of v^a on the plane (e_1, e_2)):

$$v_\alpha^a(P) = v^a(P) - (v^a(P).e_2)e_2.$$

Let us call α the true pitching angle (which is known as the angle in the plane (e_1, e_2) and the main axis of the structure). The difference between α and α_a is the angle between $V^a(P)$ and e_1 which is the direction of the true flow (in this order for the sign). It is characterized as soon as $v^a(P) \neq 0$ by:

$$\left[v_\alpha^a(P) \wedge e_1\right].e_2 = \sin(\alpha_a - \alpha)||v_\alpha^a(P)||_2, \tag{41}$$

or else (because $(v_\alpha^a(P), e_1, e_2) = v_\alpha^a(P).e_3 = v^a(P).e_3$ where $(., ., .)$ is the triple scalar product):

$$\alpha^a = \alpha + \arcsin\left(\frac{e_3.v^a(P)}{||v_\alpha^a(P)||_2}\right). \tag{42}$$

The absolute yawing angle β is the rotation around the axis e_3 of the main axis of the structure. The apparent yawing angle due to the apparent velocity is defined as follows. First of all we introduce the component of the apparent velocity in the plane (e_1, e_2) with:

$$v_\beta^a(P) = v^a - (v^a(P).e_3)e_3.$$

Hence one has (assuming that $v_\beta^a(P) \neq 0$):

$$\left[v_\beta^a(P) \wedge e_1\right].e_3 = \sin(\beta^a - \beta)||v_\beta^a(P)||_2, \tag{43}$$

or else (using the same remarks as for α^a):

$$\beta^a = \beta - \arcsin\left(\frac{e_2.v^a(P)}{||v_\beta^a(P)||_2}\right). \tag{44}$$

Finally, the last apparent angle is the rolling γ^a (γ is the nominal rolling). It is defined as the rotation around the axis e_1. The component of the apparent flow velocity in the plane (e_2, e_3) is:

$$v_\gamma^a(P) = v^a(P) - (v^a(P).e_1)e_1. \tag{45}$$

If $v_\gamma^a(P) \neq 0$ (else $\gamma = 0$) one has the relation:

$$\left[v_\gamma^a(P) \wedge e_2\right]e_1 = \sin(\gamma^a - \gamma)||v_\gamma^a||_2, \tag{46}$$

enables one to give the expression of the apparent rolling angle:

$$\gamma^a = \gamma - \arcsin\left(\frac{e_1.v^a(P)}{||v_\gamma^a(P)||_2}\right) \tag{47}$$

These three angles are chosen arbitrarily. But they are classical in many wind tunnel tests (not all). One can also use for instance the Euler angles or any other equivalent description. The change of frame is quite straightforwards. But all the angles chosen for the description of the movement, are equivalent at the first order.

Let us consider the set of the six aerodynamic or hydrodynamic coefficients at point P is denoted by c_x, c_y, c_z, c_r, c_m, c_l. The torsor of forces used in the modeling is (the coefficients are expressed in the frame (e_1, e_2, e_3) which is the reference configuration of the substructure considered but which is not fixed). We set:

$$\theta^a = (\alpha^a, \beta^a, \gamma^a).$$

Hence the aerodynamic forces are:

$$
\begin{cases}
F = \dfrac{\rho S||v^a||^2}{2}\left[c_x(\theta^a)e_1 + c_y(\theta^a)e_2 + c_z(\theta^a)e_3\right], \\[4mm]
M(P) = \dfrac{\rho SL||v^a||^2}{2}\left[c_r(\theta^a)e_1 + c_m(\theta^a)e_2 + c_l(\theta^a)e_3\right]
\end{cases}
\tag{48}
$$

4 Assembly of Several Rigid Structures

Let us turn now to the modeling of a system of several substructures. The number of substructures is N_S and we index them by the superscript $k \in \{1, 2, \ldots, N_S\}$. Each one has N_{dof}^k degrees of freedom. Some flexibility could occur for several substructures, and in this case, it would be necessary to add new degrees of freedom which would take into account this flexibility. If the movements due to this flexibility modifies the aerodynamic forces, the complexity of the model would suggest to use another modeling based on continuum mechanics. Hence let us restrict our formulation to the assembly of rigid substructures. We assume that there are connections between substructures and some of them can also be fixed to a reference frame. All these connections are taken into account in the assembling introducing stiffness matrices (K^k, C^k) for the structure S^k if it is partially or fully linked through springs to a reference frame (K^k for the displacements and C^k for the rotations) and by $(D^{k,i}, E^{k,i})$ (displacements and rotations) for the connections between the two substructures S^k and S^i. The number of neighbours of S^k and is N^k. The equations of the movement using the center of mass G of the substructure S^k are ($d^k(G)$ is the displacement of the center of mass of the kth substructure expressed in the same Galilean frame for all the substructures and r^k is the rotation of the same substructure, J_G being the inertia tensor and it should be expressed in the same basis as r^k). The aerodynamic forces are expressed in the frame linked to the substructure. Hence a change of axis is necessary if the displacements are not small enough. But if one decides to use a non Galilean frame, it could be necessary to take into account the different kind of accelerations and furthermore to modify the formulation of the links between the substructures. Hence we write formally:

$$
\begin{cases}
M\dfrac{\partial^2 d^k}{\partial t^2} + K^k d^k + \displaystyle\sum_{i=1,N^k} D^{k,i}(d^k - d^i) = F^k, \\[6mm]
J_G\dfrac{\partial^2 r^k}{\partial t^2} + C^k r^k + \displaystyle\sum_{i=1,N^k} E^{k,i}(r^k - r^i) = M^k(P^k) + F^k \wedge P^k G^k.
\end{cases}
\tag{49}
$$

The pair (d^k, r^k) is denoted by $X^k \in \mathbb{R}^6$ (three displacements and three rotations in general) in the following. But there can be some constraints which reduce the number of degrees of freedom. Hence, X^k is the vector of the degrees of freedom of the substructure S^k. The set of all the vectors X^k is denoted by X. One has $X \in \mathbb{R}^{6N_S}$ but it can be a smaller space due to the constraints mentioned previously.

For sake of convenience, the right-hand side of equations (49) is denoted by L^k. It is a vector of $\mathbb{R}^6$ (or less taking into account the constraints) for each substructure S^k which takes into account the weight, the buoyancy (for a ship) and so on. Furthermore, it can depend on X^k and $\dot{X}^k$ through the apparent angles θ^k for instance. It could also depend on the time t as far as one would take into account the instabilities of the flow which can induce buffting for instance. We also use the notation $L \in \mathbb{R}^{6N_S}$ for the set of all the vectors L^k.

The continuity relations between two adjacent substructures can be prescribed in terms of velocity continuity at each time. And from a time integration, it would imply the continuity of the kinematical variables. Therefore it can be written using a linear equation involving the velocities. It is assumed that the initial conditions satisfy these connecting conditions. Between two substructures—say S^k and S^i, the continuity of the velocity is stated by (in small displacement analysis, one can formulate the continuity on the displacements instead of the velocities):

$$B^{k,i} \dot{X}^k + B^{i,k} \dot{X}^i = 0. \tag{50}$$

The details of the matrices $B^{k,i}$ imply the description of the geometry and therefore the degrees of freedom X^k and X^i which should be updated at each instant. Let us recall that N^k is the number of substructures connected to S^k (if there is at least one!).

One can summarize the equations of the movement as follows ($\Lambda^{k,i} = \Lambda^{i,k}$ because of the reciprocity), is the Lagrange multiplier insuring the continuity of relations similar to (50); it is a function of time:

$$\begin{cases} M^k \ddot{X}^k + A^k X^k + \displaystyle\sum_{i=1,N^k} {}^t B^{k,i} \Lambda^{k,i} = L^k(X^k, \dot{X}^k), \\[2mm] \forall i = 1, N^k, \quad B^{k,i} \dot{X}^k + B^{i,k} \dot{X}^i = 0. \end{cases} \tag{51}$$

Let X_0^k be an equilibrium position (taken into account all the external forces as the weight, the buoyancy... in the expression of L^k). It is solution of the non linear system (due to $L^k(X^k, \dot{X}^k)$ and because $B^{k,i}$ are also dependent on X^k and X^i):

$$\begin{cases} \forall k = 1, N_S, \\[2mm] A^k X_0^k + {}^t B^{k,i}(X_0^k, X_0^i)\Lambda_0^{k,i} = L^k(X_0^k, 0), \\[2mm] \forall k = 1, N_S, \ \forall i = 1, N^k, \ B^{k,i} X_0^k + B^{i,k} X_0^i = 0. \end{cases} \tag{52}$$

The solution method for solving these equations is not obvious and we do not discuss it in this presentation. We assume in the following that the solution X_0^k is known for every $k = 1, N_S$.

The linearization (let us point out that the constraint (50) is satisfied by X_0) of the system close to the position X_0^k of the system leads to:

$$\begin{cases} \forall\, k = 1, N_S, \ \forall i = 1, N^k \ \Lambda^{k,i} = \Lambda^{i,k} : \\[3mm] M^k \ddot{X}^k - \dfrac{\partial L^k}{\partial \dot{X}^k}(X_0^k)\dot{X}^k + \left[A^k - \dfrac{\partial L^k}{\partial X^k}(X_0^k)\right](X^k - X_0^k) \\[4mm] \qquad + \displaystyle\sum_{i=1,N^k} {}^t B^{k,i} \Lambda^{k,i} = 0, \\[4mm] \forall i = 1, N^k, \ B^{k,i}(X^k - X_0^j) + B^{i,k}(X^i - X_0^i) = B^{k,i} X^k + B^{i,k} X^i = 0. \end{cases} \tag{53}$$

Let us underline again that the initial conditions (both the position and the velocity) satisfy the connecting conditions. A control system can be applied to some substructures. In many cases the control is precisely one or several components of the unknown X (inclination of a flap for instance). But for sake of clarity in the following, we choose to separate the two. Therefore the equations (53) are upgraded as follows. The control applied on the substructure S^k is denoted by u^k. It is possible that a given substructure has no control. In fact, it will be governed by its neighbours (for instance the main wing are often controlled through the flaps which are connected to them). At each instant $u^k \in \mathbb{R}^{q^k}$ where $q^k \in \mathbb{N}$ is the number of independent controls applied to the substructure S^k. The application to the substructure is performed through a system represented by a rectangular matrix G^k from $\mathbb{R}^{q^k}$ into $\mathbb{R}^6$. But it can contain a lot of zeros and $q^k < 6$. In most applications $q^k = 1$. This is also the case for the theoretical developments that we give in the following.

Finally, the complete linearized model close to an equilibrium point X_0^k is the following one and initial conditions should be added (anyway this is the model that we consider):

$$
\begin{cases}
\forall\, k = 1, N_S, \ \forall i = 1, N^k, \ \Lambda^{k,i} = \Lambda^{i,k} : \\[2mm]
M^k \ddot{X}^k - \dfrac{\partial L^k}{\partial \dot{X}^k}(X_0^k)\dot{X}^k + \left[A^k - \dfrac{\partial L^k}{\partial X^k}(X_0^k)\right](X^k - X_0^k) \\[4mm]
\qquad + \displaystyle\sum_{i=1,N^k} {}^t B^{k;i}\,\Lambda^{k,i} = G^k u^k, \\[4mm]
\forall i = 1, N^k, \ B^{k,i} X^k + B^{i,k} X^i = 0.
\end{cases}
\tag{54}
$$

Let us give finally a global formulation of the model including all the substructures. First of all we introduce the admissible set of state variables as follows ($B^{k,i}$ is fixed and depends on X_0):

$$
V_0 = \{ Y = \{Y^k\}_{k=1,N_S}, \ Y^k \in \mathbb{R}^6, \ \forall k = 1, N_S, \ \forall i = 1, N^k, \ B^{k,i} Y^k + B^{i,k} Y^i = 0 \}.
\tag{55}
$$

Let us point out that the initial conditions $X(0) = \{X_0^k\}$ and $\dot{X}(0) = \{X_1^k\}$ are assumed to belong to the space V_0. Then we define several matrices by:

$$
\mathcal{M} =
\begin{pmatrix}
M^1 & 0 & 0 & 0 \cdots \\
0 & M^2 & 0 & 0 \cdots \\
0 & 0 & M^3 & 0 \cdots \\
& \cdots\cdots\cdots &
\end{pmatrix}
\qquad
\mathcal{A} =
\begin{pmatrix}
A^1 - \frac{\partial L^1}{\partial X^1}(X_0) & 0 & 0 & 0\cdots \\
0 & A^2 - \frac{\partial L^2}{\partial X^2}(X_0) & 0 & 0\cdots \\
0 & 0 & A^3 - \frac{\partial L^3}{\partial X^3}(X_0) & 0\cdots \\
& \cdots\cdots\cdots\cdots &
\end{pmatrix},
$$

$$
\mathcal{G} =
\begin{pmatrix}
G^1 & 0 & 0 & 0\cdots \\
0 & G^2 & 0 & 0\cdots \\
0 & 0 & G^3 & 0\cdots \\
0 & 0 & 0 & \cdots\cdots
\end{pmatrix}
\qquad
\mathcal{D} =
\begin{pmatrix}
-\frac{\partial L^1}{\partial \dot{X}^1}(X_0) & 0 & 0 & 0\cdots \\
0 & -\frac{\partial L^2}{\partial \dot{X}^2}(X_0) & 0 & 0\cdots \\
0 & 0 & -\frac{\partial L^3}{\partial \dot{X}^3}(X_0) & 0\cdots \\
0 & 0 & 0 & 0\cdots
\end{pmatrix}
$$

$$U = \begin{pmatrix} u^1 \\ u^2 \\ u^3 \\ \cdots \end{pmatrix}, \quad \Lambda = \begin{pmatrix} \Lambda^{1,2} \\ \Lambda^{1,3} \\ \cdots \\ \Lambda^{2,3} \\ \Lambda^{2,4} \\ \cdots \end{pmatrix} \tag{56}$$

and we denote by $\mathscr{B}$ the matrix which is such that:

$$Y \in \mathbb{R}^{6N_S}, \quad \mathscr{B}Y = 0 \iff Y \in V_0. \tag{57}$$

The linearized model consists in finding:

$$\forall t > 0, \ X \in \mathbb{R}^{6N_S}, \ \Lambda \in \mathbb{R}^{\sum_{k=1,N_S} N^k}$$

satisfying the initial conditions (say $X(0)$ and $\dot{X}(0)$ given), such that:

$$\begin{cases} \mathscr{M}\ddot{X} + \mathscr{D}\dot{X} + \mathscr{A}X + {}^t\mathscr{B}\Lambda = \mathscr{G}U, \\ \\ \mathscr{B}X = 0. \end{cases} \tag{58}$$

Remark 4 One could add a right-hand side to the first equation (58). It can represent aerodynamic forces for instance. This doesn't imply any fundamental change in the theory developed hereafter as soon as we assume a sufficient regularity on F^k with respect to the time; for instance $F^k \in L^2(]0, T[; \mathbb{R}^6)$.

5 Control Strategy for the Full System

For sake of brevity, we choose a scalar function for the control function applied on each substructure (it means that $q^k = 1$). For each substructure which is equipped with a control system, we define a criterion which represents the error between the actual configuration X^k and the desired one X_d^k at a given instant—say T. It is convenient, but not necessary, to choose the same time for all the substructures. It is denoted by J^k and it is dependent on all the controls $u = \{u^k\}_{k=1,N_S}$, $u^k \in L^2(]0, T[)$ as far as the substructures are coupled.

Furthermore, if the control time T^k is defined separately for each substructure, the largest one is denoted by T and the control u^k is assumed to be zero on $]T^k, T[$ (if this set is not empty). For sake of brevity, we restrict our presentation to the case where the control is scalar on each substructure. The reader will find many improvements in the books by R. Pallu de la Barrière [25], R.E. Kalman [22], R. Bellman [4, 5] and J.L. Lions [23, 24]. Let us set for instance:

$$J^k(u^1, u^2, \ldots, u^{N_S}) = \frac{1}{2}\Big[||X^k(T^k) - X_d^k||^2 + ||\dot{X}^k(t)||^2 + \varepsilon \int_0^{T^k} |u^k(t)|^2 dt\Big].$$
(59)

The dependance of J^k with respect to u^i, $i \neq k$ is performed through the state variable X^k which is solution of global model connecting all the substructures. The most natural functional space for the control u^k in the previous formulation is $L^2(]0, T^k[)$. The control u^k applied to the substructure S^k is assumed to be a scalar function in order to simplify the writings.

It often appears that some additional regularity should be required for the exact control that we define in the following (for instance a continuous control can be more interesting than a $L^2(]0, T[)$ control. Furthermore, it is convenient in a first step to define the optimal control problem that we are considering. In a second step we discuss other possibilities. The Nash points are briefly discussed. For a precise presentation of this strategy for partial differential equations, we refer to A. Bensousan [6] and A. Bensoussan & J.L. Lions [7].

There are many strategies for defining the control laws. The first one is the cooperative optimization which is certainly the most efficient from the mathematical point of view, but not necessarily the most convenient or appropriate for the real system. The second one is the competitive strategy which consists in using, for instance, a so-called Nash equilibrium or as we do in the following, the decentralized formulation with coordination. Let us define briefly such strategies in a mathematical framework.

5.1 The Cooperative Strategy

We introduce a global criterion for the control. A compromise (or a bargaining as say the economists) attributes a relative importance to each substructure through a coefficient $a^k > 0$. Then the global criterion to be minimized is:

$$J(v) = \sum_{k=1, N_S} a^k J^k(v^k) \text{ where } v = (v^1, v^2, \ldots, v^{N_S}).$$
(60)

The global optimal control is solution of:

$$\min_{v=(v^1, \ldots, v^{N_S}) \in \Pi_{k=1, N_S} L^2(]0, T^k[)} J(v).$$
(61)

The existence and uniqueness of a solution are standard as far as the criterion J is strictly convex, continuous and coercive (tends to the infinity when the norm of the control tends also to the infinity because $\varepsilon > 0$) on the space:

$$V_{cont} = \Pi_{k=1,N_S} L^2(]0, T^k[).$$

Furthermore, if there exists at least one exact control, such that:

$$\forall k = 1, N_S, \; X(T) = \dot{X}^k(T) = 0,$$

denoted by $u^e \in V_{cont}$, the optimal control converges strongly in the space V_{cont} when $\varepsilon \to 0$, to the exact control of the space V_{cont} which has the minimum norm in this space.

It is worth noting that in this case, the only assumption required in the proof (given in the following for a simplified version; see Sect. 5.2) concerns a global controllability of the system (58). The computation of this exact limit control can be performed using the Gramian matrix as far as the exact controllability assumption is satisfied (see for instance [15]).

Nevertheless, the solution involves the full system and it depends on the choice of the coefficients a^k. This is precisely the compromise step. Details for the construction of the Gramian matrix can be found in many books. Just for sake of commodity in the reading, we summarize hereafter how this exact limit control can be computed easily.

5.2 How to Compute the Limit of the Optimal Control When $\varepsilon \to 0$

Let us consider the simple following example with only one substructure (we set $N^1 = N$, $X_d^1 = 0$ and $T^k = T$ for sake of simplicity) in order to sketch the method. The notations used in this subsection are slightly different from those of the general case. Because we have only one substructure, we also set $A^1 = A$, $M^1 = M$, $F^1 = F$, $G^1 = G$ and $X^1 = X$. Obviously, the constraint between the substructures has disappeared (ie. no coupling matrix $\mathscr{B}$). The optimal control model consists in finding $u^\varepsilon \in V_{cont}$ such that ($\varepsilon > 0$, M and A being two symmetrical and positive definite $N \times N$ matrices, G is a constant vector of $\mathbb{R}^N$ and F can be time dependent):

$$\begin{cases} \min_{v \in L^2(]0,T[)} J^\varepsilon(v) = \frac{1}{2}\{\|X(T)\|^2 + \|\dot{X}(T)\|_2^2 + \varepsilon \int_0^T |v(s)|^2 ds\} \\ \\ X(t) \in \mathbb{R}^N, \; M\ddot{X} + AX = F + Gv, \; X(0) = X_0, \; \dot{X}(0) = X_1. \end{cases} \tag{62}$$

Existence and uniqueness of a solution u^ε are standard (see for instance [11]). The optimality conditions can be formulated using the adjoint state P which is solution of:

$$M\ddot{P}^\varepsilon + AP^\varepsilon = 0, \quad MP^\varepsilon(T) = \dot{X}^\varepsilon(T), \quad M\dot{P}^\varepsilon(t) = -X^\varepsilon(T). \tag{63}$$

Hence the optimality relation is:

$$\forall t \in]0, T[, \quad \varepsilon u^\varepsilon(t) +{}^t GP^\varepsilon(t) = 0. \tag{64}$$

The idea in order to construct an exact control with a minimum cost (following the idea of Tikhonov on regularization [30]), consists in setting *a priori*:

$$\begin{cases} X^\varepsilon = X^0 + \varepsilon X^1 + \ldots, \\ P^\varepsilon = P^0 + \varepsilon P^1 + \ldots, \\ u^\varepsilon = u^0 + \varepsilon u^1 + \ldots. \end{cases} \tag{65}$$

By introducing this assumed asymptotic expansion into (62) and (64) and by equating the terms of same power in ε, one obtains necessary conditions (Φ_0, Φ_1 are unknowns at this step):

$$\begin{cases} M\ddot{X}^0 + AX^0 = F + Gu^0, \quad X^0(0) = X_0, \quad \dot{X}^0(0) = X_1, \\ M\ddot{P}^0 + AP^0 = 0, \quad MP^0(T) = \dot{X}^0(T), \quad M\dot{P}^0(T) = -X^0(T), \\ M\ddot{P}^1 + AP^1 = 0, \quad MP^1(0) = \Phi_0, \quad M\dot{P}^1(0) = \Phi_1, \\ {}^t Gp^0 = 0, \quad u^0 +{}^t GP^1 = 0. \end{cases} \tag{66}$$

It is now necessary to add the controllability hypothesis for the full system.

Controllability Hypothesis *The basic assumption is the following one (see I. Pontryaguin [26, 27] and R. Bellman [3, 4] for first order EDO and for instance [15] in the framework discussed here):*

If a vector $Z \in \mathbb{R}^N$ satisfies : ${}^t G[M^{-1}A]^i Z = 0 \; \forall i = 0, N - 1$ then $Z = 0$. $\quad\square$

$$\tag{67}$$

From ${}^t GP^0 = 0$ and ${}^t G[M^{-1}A]^i P^0(t) = 0, \; \forall t \in [0, T[, \; \forall i = 0, N - 1$ one deduces that $P^0(t) = 0$ (necessary condition). Therefore $X^0(T) = \dot{X}^0(T) = 0$.

Let us introduce $Q(t) \in \mathbb{R}^N$, $\forall t \in [0, T]$, solution of:

$$M\ddot{Q} + AQ = 0, \text{ and } Q(0) = \delta\Phi_0, \ \dot{Q}(0) = \delta\Phi_1.$$

Multiplying the equation which should be satisfied by X^0 and replacing u^0 by $-{}^tGP^1$, one obtains after two integrations by parts:

$$-M\dot{X}^0(0).\delta\Phi_0 + MX^0(0).\delta\Phi_1 = \int_0^T F(s).Q(s)ds - \int_0^T {}^tGP^1(s).{}^tGQ(s)ds.$$

Hence setting:

$$\begin{cases} \Phi = (\Phi_0, \Phi_1), \ \delta\Phi = (\delta\Phi_0, \delta\Phi_1), \ \Lambda^T(\Phi, \delta\Phi) = \int_0^T {}^tGP^1(s).{}^tGQ(s)ds, \\[2em] L^T(\delta\Phi) = \int_0^T F(s).Q(s)ds + MX_1.\delta\Phi_0 - MX_0.\delta\Phi_1, \end{cases}$$

the computation of u^0 is deduced from P^1 which is characterized from $\Phi \in \mathbb{R}^{2N}$ solution of:

$$\forall \delta\Phi \in \mathbb{R}^{2N}, \ \Lambda^T(\Phi, \delta\Phi) = L^T(\delta\Phi). \tag{68}$$

This symmetrical, linear and finite dimensional system has a unique solution because of the controllability assumption. The matrix associated to the bilinear form Λ^T is also positive definite because of the controllability hypothesis.

Once Φ is computed satisfying (68), one has conversely:

$$\forall \delta\Phi = (\delta\Phi_0, \delta\Phi_1), \ MX^0(T).\dot{Q}(T) - M\dot{X}^0(T).Q(T) = 0. \tag{69}$$

Because the mapping: $(\delta\Phi_0, \delta\Phi_1) \to (Q(T), \dot{Q}(T))$ is onto (and one to one), this implies that:

$$X^0(T) = \dot{X}^0(T) = 0,$$

which means that u^0 is perfectly defined and is an exact control. Furthermore, let us set:

$$\begin{cases} V_{ex} = \{v \in L^2(]0, T[), \text{ such that: if } Z \text{ is solution of: } M\ddot{Z} + AZ = F + Gv, \\[1em] \text{with } Z(0) = X_0, \ \dot{X}(0) = X_1, \text{ then } Z(T) = \dot{Z}(T) = 0\}. \end{cases} \tag{70}$$

The control $u^0 \in V_{ex}$ (set of exact controls) satisfies:

$\forall v \in V_{ex}$:

$$\int_0^T (u^0 v)(s)ds = -\int_0^T ({}^t GP^1.v)(s))ds = -\int_0^T (P^1.(M\ddot{Z} + AZ))(s)ds$$

$$+ \int_0^T F(s).P^1(s)ds = MP^1(0).\dot{X}_0 - M\dot{P}^1(0).X(0) + \int_0^T F(s).P^1(s)ds$$

$$= \int_0^T u^0(s)^2 ds.$$

$$(71)$$

Hence $u^0 \in V_{ex}$ satisfies:

$$\forall v \in V_{ex}, \quad \int_0^T u^0(s)(v - u^0)(s)ds = 0. \tag{72}$$

Therefore, it is the exact control with the minimum $L^2(]0, T[)$ norm and it is unique because of the strict convexity of the square of the norm.

Once u^0 is characterized, let us turn to convergence results of u^ε to u^0. Let us observe that one has the *a priori* estimate on u^ε (with respect to ε) because of its definition:

$$||X^\varepsilon(T)||^2 + ||\dot{X}(T)||^2 + \varepsilon \int_0^T |u^\varepsilon(s)|^2 ds \le \varepsilon \int_0^T |u^0(s)|^2 ds. \tag{73}$$

Consequently, one can extract a subsequence from u^ε -say $u^{\varepsilon'}$ which converges weakly to an element u^* in $L^2(]0, T[)$. Furthermore due to the continuity of the mapping $u^\varepsilon \to X^\varepsilon$, one has also $X^*(T) = \dot{X}^*(T) = 0$ (where X^* is the solution associated to u^*). Consequently $u^* \in V_{ex}$. But, from the lower semi-continuity of convex functions for the weak topology, one has:

$$\int_0^T |u^*(s)|^2 ds \le \int_0^T |u^0(s)|^2 ds. \tag{74}$$

Because of the uniqueness of the control with a minimal norm, one can conclude that $u^* = u^0$. Therefore all the sequence u^ε converges weakly to u^0 in $L^2(]0, T[)$. Concerning the strong convergence one has:

$$\int_0^T (u^\varepsilon(s) - u^0(s))^2 ds = \int_0^T (u^\varepsilon(s))^2 ds - 2 \int_0^T u^\varepsilon(s)u^0(s)ds \int_0^T (u^0(s))^2 ds$$

$$\le 2 \int_0^T u^0(s)(u^0(s) - u^\varepsilon(s))ds \to_{\varepsilon \to 0} 0. \tag{75}$$

Therefore the full sequence u^ε converges strongly in the space $L^2(]0, T[)$.

5.3 The Non-cooperative Strategy

This method is interesting in many industrial cases. Specially, if there are very different characteristic times (very different periods of vibrations) among the substructures. Some of them can be controlled by fast loops while others require slower controls. First of all, let us define the Nash optimal control in the framework that we consider in this subsection. This formulation is quite standard but the strategy based on decentralized control with coordination that we suggest, is different, up to our best knowledge. For sake of brevity, here again, we choose a scalar function for the control of each substructure (it means that $q^k = 1$). Furthermore, for sake of clarity in the notations, we prescribe here again and for the rest of the section: $T^k = T$ (same control time for all the substructures).

Definition 1 General definition of a Nash point. A control $u = (u^1, \ldots, u^{N_S}) \in V_{con} = \Pi_{k=1, N_S} L^2(]0, T[)$ is a Nash point if it satisfies:

$$\begin{cases} \forall k = 1, N_S, \ \forall v^k \in L^2(]0, T[) \\ \\ J^k(u^1, \ldots, u^{k-1}, u^k, u^{k+1}, \ldots, u^{N_S}) \le J^k(u^1, \ldots, u^{k-1}, v^k, u^{k+1}, \ldots, u^{N_S}), \end{cases} \tag{76}$$

where the criteria J^k are defined at (59) and $(X^k, \Lambda^{k,i})$ which appears in the definition of the criteria J^k, are solution of (58). □

In fact, there is another definition (see Definition 2) which is often used for exact control (the previous one is used for optimal control). It implies a coupling between the various control functions through the state variables:

Definition 2 Definition of an exact Nash equilibrium.
Let us set:

$$J_N^k(u) = \int_0^T |u^k(s)|^2 ds.$$

Let us introduce a control $u = \{u^k\}_{k=1,N_S}$, $u^k \in L^2(]0, T[)$ and $X = \{X^k\}$ the corresponding solution of (58). It is an exact control if and only if $X(T) = \dot{X}(T) = 0$. The set of exact controls for the full system described above, is denoted by E_{ex}. We say that u_N is an exact Nash equilibrium of the system (58) if and only if:

$$\begin{cases} u_N \in E_{ex}, \text{ and } \forall k = 1, N_S, \ \forall v^k \in L^2(]0, T[) \\ \\ \text{such that: } u_N = (u_N^1, \ldots \ldots, v^k, \ldots, u_N^{N_S}) \in E_{ex} \\ \\ \text{one has: } \quad J_N^k(u_N^k) \le J_N^k(v^k) \end{cases}$$

□

Remark 5 The condition prescribed in the previous definition of the set E_{ex}, implies that the solution X of the full system (58) should satisfy the condition:

$$\forall k = 1, N_S, \quad \sum_{i=1,N^k}, \quad B^{k,i} X^k + B^{i,k} X^i = 0, \quad (N^k \text{ is the number of neighbours of } S^k).$$

Therefore this relation induces a coupling between the control functions through the state variables for both Definitions 1 and 2. But the restriction is more important for the second one.   □

A third definition of non-cooperative control strategy, which is the one discussed in the following, is formulated hereafter.

Definition 3 Let $u_D = \{u_D^k\}_{k=1,N_S} \in [L^2(]0, T[)]^{N_S}$ a control and $(X, \Lambda) \in \mathbb{R}^{6N_S} \times \mathbb{R}^{\sum_{k=1,N_S} N^k}$ the corresponding solution to Eq. (58). u_D is a decentralized exact control with coordination if and only if:

• $\forall k = 1, N_S, \quad X^k(T) = \dot{X}^k(T) = 0$;

• u_D^k is the on S^k the only exact control of the space $L^2(]0, T[)$ with a minimum norm for the initial data X_0^k, X_1^k and the Lagrange multiplier Λ fixed as the solution of (58) and F^k given if one considers a given force at the right-hand side of this equation which is defined at (77).

                                                                                    □

In a first step we discuss the existence and the uniqueness of a decentralized exact control with coordination following the Definition 3. In a second step, we give an algorithm for computing this control. It is based on the method used for the proof of the existence and uniqueness of the control.

The controllability hypothesis is required for each substructure. Therefore, we assume that all the substructures equipped with a control system can be exactly controlled for any initial conditions in $\left[\mathbb{R}^{N^k}\right]^2$, any right-hand side (external force $F^k \in L^2(]0, T[; \mathbb{R}^6)$ applied to S^k) and any values of the Lagrange multiplier ensuring the connection between the substructures $\Lambda^{k,i} \in L^2(]0, T[)$, $\forall k = 1, N_S, \forall i = 1, N^k$ (let us recall that N_S is the number of substructures and N^k the number of connections with the neighbours of the substructure S^k).

5.3.1   Existence (and Uniqueness) of an Exact Decentralized Control with Coordination

It is assumed in this section, that each subsystem can be exactly controlled separately but for any initial conditions and any forces applied to it, including body forces and surface forces (Lagrange multiplier). The equations of the linearized model that we consider, are the following ones, in agreement with the formulation given at Sect. 4, where we now take into account a right-hand side

$$\mathscr{F} = \{F^k\} \in \left[L^2(]0, T[; \mathbb{R}^6)\right]^{N_S}.$$

$$\begin{cases} \text{Find } X \in \mathbb{R}^{6N_S}, \text{ and } \Lambda \in \mathbb{R}^{\Sigma_{k=1,N_S} N^k} \text{ such that:} \\[2mm] \mathscr{M}\ddot{X} + \mathscr{A}X + {}^t\mathscr{B}\Lambda = \mathscr{F} + \mathscr{G}u, \hfill (77) \\[2mm] \mathscr{B}X = 0, \quad X(0) = X_0, \quad \dot{X}(0) = X_1. \end{cases}$$

The existence and uniqueness of a solution satisfying initial conditions—say: $(X \in H^2(]0, T[, V_0), \Lambda \in [L^2(]0, T[; \mathbb{R}^{\Sigma_{k=1,N_S} N^k})$- are ensured from classical results (see for instance [20]).

Let $\Lambda = \{\Lambda^{k,i}\}$ a given multiplier corresponding to forces applied between the substructures (according to the reciprocity principle). For each substructure S^k, we define the exact control u^k considered as a function of $\Lambda^{k,i}$ as it follows.

It is the exact control at time T with the minimum $L^2(]0, T[)$-norm. We characterize it as we did in the previous Sect. 5.2.

Let us briefly sketch the method. If $\Phi_0^k = (\Phi_{00}^k, \Phi_{01}^k)$ is solution of:

$$\begin{cases} \forall \delta\Phi^k = (\delta\Phi_0^k, \delta\Phi_1^k), \ \Lambda_k^T(\Phi^k, \delta\Phi^k) = L^k(\delta\Phi^k), \\[2mm] \text{where:} \\[1mm] \Lambda_k^T(\Phi^k, \delta\Phi^k) = \int_0^T {}^tG^k P^k(s). \, {}^tG^k Q^k(s)ds, \\[2mm] L_k^T(\delta\Phi^k) = \int_0^T [F^k(s) - \sum_{i=1,N^k} {}^tB^{k,i} \Lambda^{k,i}(s)].Q^k(s)ds + M^k X_1^k.\delta\Phi_0^k - M^k X_0^k \delta\Phi_1^k, \\[2mm] \text{and:} \\[1mm] M^k \ddot{P}^k + A^k P^k = 0, \ P^k(0) = \Phi_0^k, \ \dot{P}^k(0) = \Phi_1^k, \\[2mm] M^k \ddot{Q}^k + A^k Q^k = 0, \ Q^k(0) = \delta\Phi_0^k, \ \dot{Q}^k(0) = \delta\Phi_1^k, \end{cases}$$

$$(78)$$

the control $u = (u^1, u^2, \ldots, u^k, \ldots)$ is defined on each substructure S^k by:

$$u(\Lambda) = \{u^k = -{}^tG^k P^k\}_{k=1,N_S}. \hfill (79)$$

Then we introduce $(Z, H(\Lambda))$ solution of the global model:

$$\begin{cases} \mathcal{M}\ddot{Z} + \mathcal{A}Z +{}^t\mathcal{B}H(\Lambda) = \mathcal{F} + \mathcal{G}u(\Lambda), \\ \\ \mathcal{B}Z = 0, \quad Z(0) = X_0, \quad \dot{Z}(0) = X_1. \end{cases} \tag{80}$$

Our goal is to prove that H has a fixed point—say Λ^* associated to the vector Z^* which will be solution of:

$$\begin{cases} \mathcal{M}\ddot{Z}^* + \mathcal{A}Z^* +{}^t\mathcal{B}\Lambda^* = \mathcal{F} + \mathcal{G}u(\Lambda^*), \\ \\ \mathcal{B}Z^* = 0, \quad Z^*(0) = X_0, \quad \dot{Z}^*(0) = X_1. \end{cases} \tag{81}$$

This result is obtained from a fixed point algorithm based on a contraction mapping. The consequence of this property is the uniqueness of the fixed point. This is not always the case for general Nash points for instance. One advantage of the strategy that we consider here, is to suggest a computational algorithm.

Hence, the control $u(\Lambda^*)$ is exact and locally it is the one which minimizes the $L^2(]0, T[)$-norm among the exact controls. In other words, it will be proved that it is an exact decentralized control with coordination following the Definition 3.

In order to prove the existence of a fixed point for the mapping H introduce at (80), we suggest to prove that it is a contraction on the space $[L^2(]0, T[]^{\sum_{k=1,N_S} N^k}$ (functional space for the multiplier).

Let set construct the sequence $\xi^{n+1} = H(\xi^n) \in L^2(]0, T[]^{\sum_{k=1,N_S} N^k}$ and let us set:

$$\underline{\xi}^n = \xi^{n+1} - \xi^n, \quad \underline{Z}^n = Z^{n+1} - Z^n$$

where (Z^{n+1}, ξ^{n+1}) is the solution of:

$$\begin{cases} \mathcal{M}\ddot{Z}^{n+1} + \mathcal{A}Z^{n+1} +{}^t\mathcal{B}\xi^{n+1} = \mathcal{F} + \mathcal{G}u(\xi^n), \\ \\ \mathcal{B}Z^{n+1} = 0, \quad Z^{n+1}(0) = X_0, \quad \dot{Z}^{n+1}(0) = X_1. \end{cases} \tag{82}$$

One has the following characterization (we set $\underline{\delta}^n = u(\xi^n) - u(\xi^{n-1})$):

$$\begin{cases} \mathcal{M}\ddot{\underline{Z}}^n + \mathcal{A}\underline{Z}^n +{}^t\mathcal{B}\underline{\xi}^n = \mathcal{G}(u(\xi^n) - u(\xi^{n-1})) = \mathcal{G}\underline{\delta}^n, \\ \\ \mathcal{B}\underline{Z}^n = 0, \quad \underline{Z}^n(0) = 0, \quad \dot{\underline{Z}}^n(0) = 0. \end{cases} \tag{83}$$

The term $\underline{\delta}^n = \{(\underline{\delta}^k)^n\}$ is solution of the following system:

$$
\begin{cases}
\text{Let us introduce } P(\underline{\xi}^{n-1}) = \{P^k(\underline{\xi}^{n-1})\} \text{ which is the unique solution of:} \\[2mm]
M^k \ddot{P}^k(\underline{\xi}^{n-1}) + A^k P^k(\underline{\xi}^{n-1}) = 0, \; P^k(\underline{\xi}^{n-1})(0) = \Phi_0^k, \; \dot{P}^k(\underline{\xi}^{n-1})(0) = \Phi_1^k, \\[2mm]
\text{where } \Phi^k = (\Phi_0^k, \Phi_1^k) \in \mathbb{R}^{2N^k} \text{ is the solution of:} \\[2mm]
\forall \delta \Phi^k = (\delta \Phi_0, \delta \Phi_1^k), \quad \Lambda^T(\Phi^k, \delta \Phi^k) = -\int_0^T {}^t B^k \underline{\xi}^{n-1}(s) Q(s) ds, \\[2mm]
\text{with: } {}^t B^k \xi = \sum_{i=1,N^k} {}^t B^{k;i} \xi^{k,i}, \quad \text{and} \\[2mm]
M^k \ddot{Q}^k + A^k Q^k = 0, \; Q^k(0) = \delta \Phi_0^k, \; \dot{Q}^k(0) = \delta \Phi_1^k. \\[2mm]
\text{Finally:} \\[2mm]
(\underline{\delta}^k)^n = -{}^t G^k P^k(\underline{\xi}^{n-1}) = u^k(\xi^n) - u^k(\xi^{n-1}).
\end{cases}
$$

$$(84)$$

The question to be solved now is to prove that H is a contraction. This is the goal of the next Theorem.

Theorem 1 *We assume that the local controllability of the system is satisfied and that T is small enough. Then there exists a constant $\eta \in]0, 1[$ depending on the data such that:*

ξ^n, ξ^{n-1} *being defined above*, $\|H(\xi^n) - H(\xi^{n-1})\|_{0,]0,T[} < \eta \|\xi^n - \xi^{n-1}\|_{0,]0,T[}.$

$\square$

Remark 6 Once this result proved, we can state that the sequence $\xi^{n+1} = H(\xi^n)$ is necessarily a Cauchy sequence in the space $L^2(]0, T[)^{\sum_{i=1,N_S} N^k}$ and therefore converges to a an element ξ^* solution of $\xi^* = H(\xi^*)$. Finally, it will prove the existence and uniqueness of a solution to (81) where u is an exact control such that u^k is locally an exact control with a minimal norm. But $\{u^k\}$ is also a global exact control.

The uniqueness of the limit and therefore of a decentralized exact control in our framework, could be deduced from the contraction property of H. In fact, if there were two solutions—say ξ_1^* and ξ_2^* one would have:

$$\|\xi_1^* - \xi_2^*\|_{0,]0,T[} = \|H(\xi_1^*) - H(\xi_2^*)\|_{0,]0,T[} < \|\xi_1^* - \xi_2^*\|_{0,]0,T[},$$

which implies $\xi_1^* = \xi_2^*$.

$\square$

Proof of Theorem 1 In order to derive an upper-bound on the norm of the mapping H, we need two estimates.

- **First estimate on $\underline{Z}^n$.**
  Multiplying Eq. (83) by $\underline{\dot{Z}}^n$ one obtains:

$$\frac{d}{dt}[(\mathcal{M}\underline{\dot{Z}}^n.\underline{\dot{Z}}^n) + (\mathcal{A}\underline{Z}^n.\underline{Z}^n)] = 2(\mathcal{G}\delta^n.\underline{\dot{Z}}^n) = -2\mathcal{G}^t\mathcal{G}P(\underline{\xi}^{n-1}).\underline{\dot{Z}}^n \qquad (85)$$

Or else, for any $\alpha > 0$, using Cauchy-Schwarz triangular inequality and by integrating from 0 to T (let us recall that $\underline{Z}^n(0) = \underline{\dot{Z}}^n(0) = 0$ and c_1 denoting a constant depending on the model but neither on n nor t):

$$(\mathcal{M}\underline{\dot{Z}}^n.\underline{\dot{Z}}^n)(t) + (\mathcal{A}\underline{Z}^n.\underline{Z}^n)(t) \leq$$
$$\alpha \int_0^t |\underline{\dot{Z}}^n|^2(s)ds + \frac{c_1}{\alpha}\int_0^t |{}^t\mathcal{G}P(\underline{\xi}^{n-1})|^2(s)ds. \qquad (86)$$

Let us recall that $\mathcal{M}$ is a positive definite matrix. Hence there is a strictly positive constant c_0 such that:

$$c_0|\underline{\dot{Z}}^n|^2(t) - \alpha\int_0^t |\underline{\dot{Z}}^n|^2(s)ds \leq \frac{c_1}{\alpha}\int_0^t |{}^t\mathcal{G}P(\underline{\xi}^{n-1})|^2(s)ds. \qquad (87)$$

Or else, using Gronwall's Lemma and the definition of $P(\underline{\xi}^{n-1})$:

$$\int_0^t |\underline{\dot{Z}}^n|^2(s)ds \leq [\frac{c_1}{\alpha^2}\int_0^T |{}^t\mathcal{B}\underline{\xi}^{n-1}|^2(s)ds](e^{\frac{\alpha}{c_0}t} - 1). \qquad (88)$$

Therefore:

$$c_0|\underline{\dot{Z}}^n|^2(t) \leq \frac{c_1}{\alpha}e^{\frac{\alpha t}{c_0}}[\int_0^T |{}^t\mathcal{B}\underline{\xi}^{n-1}|^2(s)ds]. \qquad (89)$$

Using the Poincaré's Lemma ($\underline{Z}^n(0) = 0$):

$$\int_0^T |\underline{Z}^n(s)|^2ds \leq T^2\int_0^T |\underline{\dot{Z}}_n(s)|^2ds,$$

and noting the positivity of $\mathcal{A}$:

$$\int_0^T |\underline{\dot{Z}}^n(s)|^2ds + \frac{1}{T^2}\int_0^T |\underline{Z}^n(s)|^2ds \leq \frac{2Tc_1}{c_0\alpha}e^{\frac{\alpha T}{c_0}}[\int_0^T |{}^t\mathcal{B}\underline{\xi}^{n-1}|^2(s)ds]. \qquad (90)$$

- **Estimate on** $\xi^{n+1} - \xi^n = H(\xi^n) - H(\xi^{n-1})$

One can consider, after a change of variables, that $\mathcal{M} = I_d$. Hence, by multiplying Eq. (83) by $\mathcal{B}$ (let us underline that $\mathcal{B}\ddot{\underline{Z}}^n = 0$) and post-multiplying by $\underline{\xi}^n$, one obtains:

$$|{}^t\mathcal{B}\underline{\xi}^n|^2(t) = -\mathcal{A}\underline{Z}^n(t).{}^t\mathcal{B}\underline{\xi}^n(t) - \mathcal{G}^t\mathcal{G}P(\underline{\xi}^{n-1}(t)).{}^t\mathcal{B}\underline{\xi}^n(t), \tag{91}$$

which leads to (c_2 being a new constant including the largest eigenvalue of $\mathcal{A}$ and depending also on $\mathcal{G}$):

$$\int_0^T |{}^t\mathcal{B}\underline{\xi}^n|^2(s)ds \leq$$

$$c_2 \int_0^T |{}^t\mathcal{B}\underline{\xi}^n|^2(s)ds]^{1/2} \{ \int_0^T |\underline{Z}^n|^2(s)ds]^{1/2} + [\int_0^T |P(\underline{\xi}^{n-1})|^2(s)ds]^{1/2}\}. \tag{92}$$

But $P(\underline{\xi}^{n-1})$ (in fact each component $P^k(\underline{\xi}^{n-1})$) is solution of a differential equation with a zero right-hand side. Therefore one has the classical energy estimate with respect to the initial conditions ($\mathcal{M}$ has been chosen equal to the identity for sake of simplicity and c_{00} is a constant independent on neither Φ_0 nor Φ_1):

$$\forall t > 0, \ ||\dot{P}(\underline{\xi}^{n-1})(t)||^2 + (\mathcal{A}P(\underline{\xi}^{n-1})(t).P((\underline{\xi}^{n-1}))(t)) \leq c_{00}(||\Phi_0||^2 + ||\Phi_1||^2).$$

Furthermore, $\Phi = (\Phi_0, \Phi_1)$ is solution of (84) and satisfies (c_3 is a constant depending on the bilinear form Λ and on $\mathcal{B}$):

$$|\Phi_0|^2 + |\Phi_1|^2 \leq c_3 \Big(\int_0^T |{}^t\mathcal{B}\underline{\xi}^{n-1}(s)|^2 ds \Big).$$

Therefore, ($\mathcal{A}$ is assumed to be positively definite and c_4 being another constant):

$$\int_0^T [|\dot{P}(\underline{\xi}^{n-1})(s)|^2 + |P(\underline{\xi}^{n-1})(s)|^2]ds \leq c_4 T \int_0^T |\underline{\xi}^{n-1}(s)|^2 ds.$$

This enables one to derive the estimate (c_5 is another constant):

$$[\int |{}^t\mathcal{B}\underline{\xi}^n(s)|^2 ds]^{1/2} \leq c_5 \sqrt{T}(1 + \frac{2c_1 T^2}{c_0\alpha} e^{\frac{\alpha T}{2c_0}})]^{1/2} [\int_0^T |\underline{\xi}^{n-1}|^2(s)ds]^{1/2}. \tag{93}$$

Let us recall that $\mathscr{B}$ is onto. Therefore ${}^t\mathscr{B}$ is one to one (hence $\mathscr{B}{}^t\mathscr{B}$ is symmetrical and positively definite with a smallest eigenvalue c_6):

$$[\int_0^T |\underline{\xi}^n|^2(s)ds]^{1/2} \le \frac{c_5}{\sqrt{c_6}} \sqrt{T}(1 + \frac{2c_1 T^2}{c_0\alpha} e^{\frac{\alpha T}{2c_0}})^{1/2}[\int_0^T |\underline{\xi}^{n-1}|^2(s)ds]^{1/2},$$
(94)

which proves Theorem 1 for T small enough so that:

$$\frac{c_5}{\sqrt{c_6}} \sqrt{T}(1 + \frac{2c_1 T^2}{c_0\alpha} e^{\frac{\alpha T}{2c_0}})^{1/2} < 1.$$

The result is true for any $\alpha > 0$ but one could choose the value which maximizes the upper bound for T.   □

Remark 7 We proved in the previous result, that the global system admits an exact control even if the hypothesis $\mathscr{H}_1$ is not required. But hypothesis $\mathscr{H}_2$ is necessary. Hence this second hypothesis implies that there exists a unique exact global control as soon as T is small enough. One could say that there exists another exact global control (assuming $\mathscr{H}_1$ with the only control on S^k) with the same data—say $u_G = \{u_G^k\}$—which is obtained by minimizing globally the criterion $J^k(v)$, over the set $v \in E_{ex}$ for k fixed, where all the other components (different from u_G^k) of the control are fixed and equal to the other values of the exact control that we have found with the fixed point of the mapping H and denoted by u_D. It is an exact decentralized control with coordination. Hence one would have:

$$\forall v = \{v_G^k\}, \quad J^k(u_G) = \int_0^T |u_G^k|^2(s)ds \le J^k(v) = \int_0^T |v^k|^2(s)ds.$$

One can notice that (only the control on S^k is changed in the definition of u_G):

$$u_G = \{u_D^1, u_D^2, \ldots, u_G^k, \ldots, u_D^{N_S}).$$

Hence one has $J^k(u_G^k) \le J^k(u_D^k)$.

Therefore, one can claim that:

- u_G^k is the minimum exact control in $L^2(]0, T[)$-norm for S^k;
- unless u_D is a Nash exact control (see Definition 2), $u_G \ne u_D$.

□

Remark 8 Each component of the decentralized control is locally exact; hence because of the continuity of the control with respect to Λ, the limit control is also exact and the solution X satisfies the continuity condition at the interfaces between the structures. Hence after convergence, the decentralized control is an exact control for each substructure which has the minimum $L^2(]0, T[)$-norm as far as the data F^k, X_0^k, X_1^k and $\Lambda^{k,j}$ are given (for each component u^k). This local control takes into account the best choices of the other substructures. But it is not a global HUM control because it doesn't minimize a unique criterion as defined in Sect. 5.1 at (60). This suggests to compute also the global HUM control for several *bargaining* between the various combination of the local criteria, in order to have estimates on lower bounds of the cost of the control. Obviously, many different strategies could be defined from these results.   □

Remark 9 In order to have fast and more efficient control, one can use another norm for the cost of the control. For instance, one could set, instead of $J^k(v^k)$ defined at (59):

$$J_{\zeta^k}^k(v^k) = \int_0^T e^{-2\zeta^k t} |v^k|^2(s) ds. \tag{95}$$

For ζ^k large enough this prescribes the control to be more efficient near $t = 0$. In fact, the value of ζ^k can be adjusted with respect to the stiffness of the substructure considered (ζ^k larger for stiffer substructures for instance). It is also possible to choose different weight functions and different time T for each substructure.   □

Remark 10 One could object that the restriction on the time T is due a technical problem in the estimates. This is surprising as far as in PDEs the restriction on time T is that it should be large enough. But we are in a finite dimensional case for which there is no lower bound on T as far as there is no restriction on the amplitude of the control.

Nevertheless the exponential term appearing in the upper bound on T, is connected to the algorithm that we are using. Therefore it seems necessary to restrict the time delay in order to avoid a *capharnaüm* which would occur if the informations have the time to travel too much from one substructure to the other. This is a strange conclusion which consists to say that a non-cooperative equilibrium can be obtained if one don't let too much time to the *discussion* between the substructures.   □

5.4 A Coordination Algorithm for the Decentralized Control

The method used in the proof of Theorem 1 is a fixed point method based on a contraction mapping. It can be applied numerically in order to compute the decentralized equilibrium with coordination. Furthermore, its convergence is geometric as far as T is small enough, but not too much in order to avoid large amplitude of the control. Thus, one possible algorithm is the following one:

An Algorithm for computing an exact decentralized control of the model considered.

1. **Step 1** Computation of the exact local control $u_0^k = -{}^t G^k P_0^k$ corresponding to the local HUM control without interaction between the substructures ($\{A^{k,i}\} = 0$). It takes into account the initial conditions on S^k and the right-hand side $\{F^k\}_{k=1,N_S}$.

2. **Step 2** Computation of a global field X^0 satisfying the continuity condition $\mathscr{B}X^0 = 0$ with the right-hand side on each substructure, the control $G^k u_0^k$ and the initial conditions. The Lagrange multiplier of the continuity condition is Λ^0.

3. **Step3** Computation of a new local control $\{u_1^k\}_{k=1,N_S}$ taking into account the edge pressure Λ^0, the right-hand side $\{F^k\}_{k=1,N_S}$ and the initial conditions.

4. **Step 4** Computation of X^1 and the Lagrange multiplier Λ^1 with the right-hand side F^k, the control vector $\{G^k u_1^k\}_{k=1,N_S}$ and still the same initial conditions (X_0, X_1).

5. **Step5** Assuming that $\Lambda^n = \{\Lambda^{k,j}\}^n$ is known, computation of the local exact control $u_n = \{u_n^k\}_{k=1,N_S}$,

6. **Step 6** Computation of X^{n+1} and Λ^{n+1} with the right-hand side $\{F^k\}_{k=1,N_S}$, $\{G^k u_{n+1}^k\}_{k=1,N_S}$ and the same initial conditions.

7. **Step 7** Convergence test on $\|\Lambda^{n+1} - \Lambda^n\|_{0,]0,T[}$.

8. **Step 8** Stop if the test is satisfied or go to a next iteration if it is not.

Remark 11 The control time T should be small enough in order to ensure that H is contractant. The upper bound for this term (maximum control time) can be estimated numerically. $\qquad\qquad\square$

Remark 12 The existence and uniqueness of an exact decentralized control as introduced at Definition 3, are new results for the kind of problems that we consider in this text. Furthermore, the fact that the computational algorithm implies a contraction mapping gives an interesting property for the application. $\qquad\square$

6 Limit Cycle of Oscillations

When an instability occurs, the engineers try to detect the possibility of a limit cycle of oscillation which is a non-linear vibration. Let us start with some basic concepts.

6.1 Definition of a Limit Cycle of Oscillations

Let us consider a non linear single degree of freedom model:

$$\ddot{x} + \omega^2 x = f(x, \dot{x}), \quad x(0) = x_0, \quad \dot{x}(0) = x_1. \tag{96}$$

The existence and uniqueness of a solution are classical for any finite time as far as the function f is—for instance—$\mathscr{C}^0(\mathbb{R}^2)$ (see E. Cartan [10]). For sake of brevity, we assume that $f(0, 0) = 0$. We also consider that $f \in \mathscr{C}^2([0, T])$ and we set:

$$\omega^2 \xi = \frac{\partial f}{\partial \dot{x}}(0, 0) \text{ and } \omega^2 v = \frac{\partial f}{\partial x}(0, 0), \quad g(x, \dot{x}) = f(x, \dot{x}) - \omega^2 \xi \dot{x} - \omega^2 v x. \tag{97}$$

Hence the simple model becomes:

$$\ddot{x} - \omega^2 \xi \dot{x} + \omega^2 (1 - \upsilon)x = g(x, \dot{x}), \qquad (98)$$

and in the neighborhood of the origin, the function g satisfies (c is a constant only dependent on the size of the neighborhood and on g):

$$|g(x, \dot{x})| \le c(|x|^2 + |\dot{x}|^2).$$

Definition 4 Limit cycle of oscillation. A solution $x(t)$ of Eq. (98) which not identically zero, is a limit cycle of oscillations if there exists a finite number $T > 0$ such that:

$$\forall t \ge 0, \ x(t + T) = x(t).$$

T is the period of the cycle. One has also $\forall t \ge 0, \ \dot{x}(t + T) = \dot{x}(t)$. In the phase plane $((x, \dot{x})$ the curve representing the function:

$$t \in [0; T] \rightarrow (x(t), \dot{x}(t)),$$

is closed. □

The first question to be discussed is the localisation of a possible limit cycle of oscillations (sometimes denoted by lco in the following).

6.2 Localisation of Limit Cycles of Oscillations

Let us mention two criteria which are often used by the engineers. The first is due to Poincaré and Bendixson and the second one is based on the energy balance during a lco.

Theorem 2 (Poincaré-Bendixson Criterion) *Let us assume that Eq. (96) admits a lco with period T. The curve representing the limit cycle in the phase plane is denoted by $\mathscr{C}$ and the domain delimited by it is $\mathscr{D}$. Let us define two open sets in the phase plane:*

$$\mathscr{B}^+ = \{(x_1, x_2) \in \mathbb{R}^2, \ \frac{\partial f}{\partial \dot{x}}(x_1, x_2) > 0\},$$

$$\mathscr{B}^- = \{(x_1, x_2) \in \mathbb{R}^2, \ \frac{\partial f}{\partial \dot{x}}(x_1, x_2) < 0\}$$

Then the domain $\mathscr{D}$ delimited by the lco can't be included neither in $\mathscr{B}^+$ nor in $\mathscr{B}^-$.

□

Proof of Theorem 2 In the phase plane, the lco is a closed curve denoted by $\mathscr{C}$. The unit normal to this curve and oriented outwards of the domain $\mathscr{D}$ delimited by $\mathscr{C}$ is denoted by ν. The tangent to the curve $\mathscr{C}$ is parallel to the vector : $p = (\dot{x}, \ddot{x}) = (\dot{x}, f(x, \dot{x}) - \omega^2 x)$. Hence along this curve, one has: $p.\nu = \nu_1 \dot{x} + \nu_2(f(x, \dot{x}) - \omega^2 x) = 0$. Or else, using Stokes formula:

$$\int_{\mathscr{C}} p.\nu = 0 = \int_{\mathscr{D}} \text{div } (p) = \int_{\mathscr{D}} \frac{\partial f}{\partial \dot{x}}(x_1, x_2) \ (= \int_{\mathscr{D}} [\frac{\partial g}{\partial \dot{x}}(x_1, x_2) + \omega^2 \xi]).$$

Therefore $\mathscr{D}$ can't be in neither $\mathscr{B}^+$ nor in $\mathscr{B}^-$.   □

Remark 13 It is worth to point out that the Poincaré-Bendixson criterion only gives informations on the domain delimited by a lco. It is possible that the curve $\mathscr{C}$ could be included in $\mathscr{B}^+$ or $\mathscr{B}^-$.   □

Remark 14 The Poincaré-Bendixson criterion doesn't give any information on the existence of a lco. But it can be used in order to prove that there is no lco for a given dynamical system. For instance, in the case where $\mathscr{B}^+ = \mathbb{R}^2$.   □

Theorem 3 (Localisation of a lco Using the Energy) *Let us consider here again that the system (96) has a lco which is represented in the phase plane by a closed curve $\mathscr{C}$. We introduce the function:*

$$h(x_1, x_2) = \frac{f(x_1, x_2) - f(x_1, 0)}{x_2}.$$

$$\mathscr{E}^+ = \{(x_1, x_2) \in \mathbb{R}^2, \ h(x_1, x_2) > 0\},$$

$$\mathscr{E}^- = \{(x_1, x_2) \in \mathbb{R}^2, \ h(x_1, x_2) < 0\}.$$

The result of the Theorem is that the curve $\mathscr{C}$ can't be included neither in $\mathscr{E}^+$ nor $\mathscr{E}^-$.   □

Proof Let us multiply Eq. (96) by $\dot{x}$ assuming that the solution $x(t)$ is a lco. By integrating this expression from 0 to T (the period of the lco), one obtains, because of the periodicity of x and because $\int_0^T f(x, 0)\dot{x} = 0$:

$$\int_0^T f(x, \dot{x})\dot{x} = \int_0^T [f(x, \dot{x}) - f(x, 0)]\dot{x} = \int_0^T [\frac{f(x, \dot{x}) - f(x, 0)}{\dot{x}}]\dot{x}^2 = 0. \tag{99}$$

Hence the lco can't be included neither in $\mathscr{E}^+$ nor $\mathscr{E}^-$.   □

Remark 15 It is worth noting that in Theorem 3 the condition concerns the trajectory of the lco. But, in case of Theorem 2, it concerns the domain delimited by the trajectory.   □

Remark 16 Let us notice that:

$$h(x, \dot{x}) = \frac{f(x, \dot{x}) - f(x, 0)}{\dot{x}} = \frac{g(x, \dot{x}) - g(x, 0)}{\dot{x}}.$$

This enables one to compare the informations deduced from the two criteria on some simple examples. □

Remark 17 Let consider a simple example where the function f can be estimated by a cubic as follows:

$$f(x, \dot{x}) = Ax + B\dot{x} + Cx^2 + Dx\dot{x} + E\dot{x}^2 + Fx^3 + Gx^2\dot{x} + Hx\dot{x}^2 + I\dot{x}^3.$$

Then one has:

$$\begin{cases} \dfrac{\partial f}{\partial \dot{x}}(x, \dot{x}) = B + Dx + 2E\dot{x} + Gx^2 + 2Hx\dot{x} + 3I\dot{x}^2, \\[2mm] \text{and} \\[2mm] h(x, \dot{x}) = B + Dx + E\dot{x} + Gx^2 + Hx\dot{x} + I\dot{x}^2. \end{cases}$$

Let us consider the case where $B < 0$ (negative damping as in stall flutter phenomenon studied in Sects. 2.6–2.9 and 8. The origin is in the set $\mathscr{B}^-$ and the system is unstable around the origin. The set $\mathscr{B}^-$ is delimited by an ellipse curve. The most interesting case is the one of an ellipse ($H^2 - 3GI < 0$ which implies that $GI > 0$). The Poincaré-Bendixson criterion states that a lco (if there is one), should cross this ellipse. In a similar way the energy criterion indicates the same thing but with another curve, the equation of which is:

$$B + Dx + E\dot{x} + Gx^2 + Hx\dot{x} + I\dot{x}^2. = 0.$$

Because of the former condition $H^2 - 3GI < 0$, one can claim that this latter curve is also an ellipse ($H^2 - 4GI < H^2 - 3GI < 0$). □

Let us go now to the existence (but not uniqueness) of a lco for Eq. (98). First of all, it is necessary to define invariant sets by Eq. (98).

Definition 5 Invariant sets by an EDO. Let $\mathscr{I}$ a subset of the phase plane. It is said to be invariant by Eq. (98) if for any initial condition in $\mathscr{I}$, and for the solution $x(t)$ satisfying these initial condition, one has:

$$\forall t \geq 0, \ (x(t), \dot{x}(t)) \in \mathscr{I}.$$

□

Theorem 4 *Let us assume that there exists a compact invariant set $\mathscr{I}$ for Eq. (98) and the function f is assumed to be at least $\mathscr{C}^2(\mathbb{R}^2)$ (not necessary but convenient). Then one of the two situations occurs (may be the two):*

1. *there is a limit point on the horizontal axis of the phase plane (i.e. $\lim_{t\to\infty} x(t) = x_s$ where x_s is solution—if there is one in $\mathscr{I}$—of the equation: $\omega^2 x_s = f(x_s, 0)$);*
2. *there is, at least, one limit cycle in $\mathscr{I}$ (with a finite period T).*

$\square$

Proof Let us consider an arbitrary point (x_1, x_2) in $\mathscr{I}$. Let us choose for instance $x_2 > 0$. The solution of (98) with this initial condition is denoted by $x(t)$. Two situations can occur:

1. there exists a finite time t_1 for which $x(t_1) = 0$. There are also two possibilities:

   - (a) first, if $x(t_1)$ is solution of the static equilibrium equation:

     $$\omega^2 x(t_1) = f(x(t_1), 0),$$

     the Theorem 4 is proved;
   - (b) second, if $x(t_1)$ is not a solution of the static equation, the trajectory representing the solution crosses the axis $\dot{x} = 0$. and we have the same discussion as before but in the half plane $\dot{x} \leq 0$.

2. $\forall t \geq 0$ one has $\dot{x}(t) > 0$. Because the solution remains bounded ($\mathscr{I}$ is compact), and is increasing ($\dot{x} > 0$), there is a limit to $x(t)$ when $t \to \infty$, say x^*. Obviously $\lim_{t\to\infty} \dot{x}(t) = 0$. Hence the point $(x^*, 0)$ is on the axis $\dot{x} = 0$. It remains to prove that $\lim_{t\to\infty} \ddot{x}(t) = 0$ in order to complete the proof of Theorem (4) in this second case. But from (98) one can claim that $\ddot{x}(t)$ has a limit when $t \to \infty$ which is equal to $c_0 = f(x^*, 0) - \omega^2 x^*$. Let us assume that $c_0 > 0$ for instance. For $(t_2 > t_1)$ one has:

$$\dot{x}(t_2) - \dot{x}(t_1) = \int_{t_1}^{t_2} \ddot{x}(s)ds.$$

Let $\varepsilon > 0$ and $t_1(< t_2)$ large enough. One has:

$$|\dot{x}(t_2) - \dot{x}(t_1)| \geq (c_0 - \varepsilon)(t_2 - t_1) > 0.$$

Setting $t_2 = t_1 + 1$ and by taking the limit when $t_1 \to \infty$, we obtain:

$$c_0 < 0$$

which requires that $c_0 = 0$. The same proof could be given for $c_0 < 0$. Hence the Theorem 4 is proved in this case.

It remains to consider the situation where one can define an infinite sequence of instants t_{2n}, $n \in \mathbb{N}^*$ (the trajectory comes from the bottom) and t_{2n-1} (trajectory comes from the top) corresponding to intersection of the trajectory in the phase plane with the axis $\dot{x} = 0$.

Let us consider for instance the sequence $x(t_{2n-1})$. Because of the uniqueness of the solution for any compact interval in time, the trajectories can't cross each other. Therefore the sequence $x(t_{2n-1})$ is monotonic (increasing or decreasing). Because the solution remains in a compact, it is bounded and thus it converges to a point say x^∞. The solution of (96) initiated at the point $(x^\infty, 0)$ is a limit cycle of oscillations. This completes the proof of Theorem 4. □

Remark 18 Even if this Theorem 4 seems to be attractive, the main difficulty remains to find invariant sets. This is partially treated in the next subsection. □

6.3 How to Find Invariant Sets?

There is no general method for determining invariant sets which guaranty the existence of a lco. But some computational tricks which are strongly dependent on the expression of the function f can be used. These tricks are mainly based on changes of variables (which will be discussed in the next session) and multiplier technics which are basically connected to the search of invariant quantities to Eq. (96).

The first attempt consists in multiplying this equation by $\dot{x}$. This leads to (energy invariant):

$$\frac{d}{dt}[\dot{x}^2 + \omega^2 x^2] = 2f(x, \dot{x})\dot{x}. \tag{100}$$

Let us consider the following expression $f(x, \dot{x}) = p(x) + \dot{x}q(x, \dot{x})$. Anyway it is a quite general formulae derived from a third order Taylor expansion of f. Let P be the primitive of p equal to zero for $x = 0$ and $q = a + bx + c\dot{x} + dx^2 + ex\dot{x} + f\dot{x}^2$. Hence one has:

$$f(x, \dot{x})\dot{x} = p(x(t))\dot{x}(t) + q(x(t), \dot{x}(t))\dot{x}^2(t)$$

$$= \frac{d}{dt}[P(x)] + \dot{x}^2[a + bx + c\dot{x} + dx^2 + ex\dot{x} + f\dot{x}^2].$$

Let us introduce the curves in the phase plane defined by the equation:

$$\dot{x}^2(t) + \omega^2 x^2 - 2P(x) = \text{constant}, \tag{101}$$

and let us assume that it is a close curve (this only depends on $p(x)$). Furthermore we introduce the domain $\mathscr{K}$ delimited by a conic:

$$\mathscr{K} = \{(x, \dot{x}) \in \mathbb{R}^2, \quad a + bx + c\dot{x} + dx^2 + ex\dot{x} + f\dot{x}^2 \leq 0\}. \tag{102}$$

In the case of a stall flutter (see Sects. 2.6 and 2.9), one has $a > 0$. Let us assume for instance that the curve defining the boundary of $\mathscr{K}$ is an ellipse (which contains the origin). Hence, the domain $\mathscr{K}$ is the outside of the conic because $a > 0$.

Let us now consider the quantity:

$$t > 0 \rightarrow \dot{x}^2 + \omega^2 x^2 - 2P(x), \tag{103}$$

where x is a solution of Eq. (96) (which is unstable in a close neighborhood of the origin because $a > 0$ (and this implies a negative damping)). The curve defined by:

$$\dot{x}^2 + \omega^2 x^2 - 2F(x) = \delta_c, \tag{104}$$

is closed as far as (for instance, but not only) $F(x) < \omega^2 x^2$. When it remains inside the set delimited by the conic $\mathscr{E}$, it is increasing (δ_c increases) until it reaches the conic $\mathscr{E}$ delimiting the set $\mathscr{K}$ at a given point $c = (x_c, \dot{x}_c)$ and at a finite time $t > 0$.

From this point, there is a unique closed curve (assumption) corresponding to Eq. (104). Because the mapping $\delta_c \rightarrow$ *the curve*, is continuous, and because the values of δ_c are bounded when the point c describes the conic (assumed to be an ellipse) $\mathscr{E}$, this enables one to claim that δ_c reaches its upper bound.

From the relation (100), one can ensure that the part of the curve in $\mathscr{K}$ corresponding to the expression $\dot{x}^2 + \omega^2 x^2 - 2F(x)$, where x is the solution of (96) remains inside the one defined at (110) until it comes back into the interior of the ellipse $\mathscr{E}$ defining the boundary of $\mathscr{K}$. Let us define by $\mathscr{P}$, the external envelope of all the curves (104) for $c \in \mathscr{E}$. The pair $(x, \dot{x})$ solution of (96) initiated inside $\mathscr{P}$ remains in this set. Therefore it is an invariant set through Eq. (96). Let us give a simple example with: $q = a - (x^2 + \dot{x}^2)$ where $a > 0$, $\omega = 1$ and $p = 0$. One has an instability in the vicinity of the origin (negative damping). The boundary of $\mathscr{K}$ is a circle defined by:

$$\mathscr{E} = \{(x, \dot{x}) \in \mathbb{R}^2, \ x^2 + \dot{x}^2 = a\}. \tag{105}$$

The curves $\mathscr{P}$ is the same as $\mathscr{E}$. Other examples are plotted on Fig. 22 for $\omega \neq 1$.

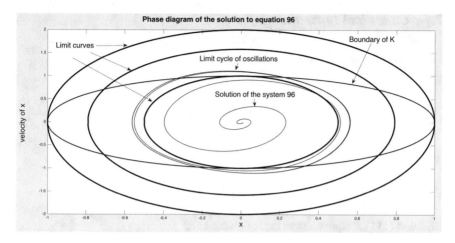

Fig. 22 This plot represents four different curves: (i) the solution $x(t)$ of the equation (96); (ii) the conic $\mathscr{E}$ delimiting the set $\mathscr{K}$; (iii) three different limit curves the envelope of which defines the invariant set; (iv) the limit cycle of oscillations of the solution of (96)

7 Computation of the Limit Cycle

The idea was introduced by different authors. Certainly, the most important contribution is due to V.I. Arnold [1], but several very nice presentations have been suggested in many books. Let us mention the one of S. Wiggins [31] who gave a simple and clear presentation.

The method is called the normal form algorithm. It is based on a sequence of changes of variables. Let us give a short description of the method on the scalar equation where $a > 0$ is a negative damping as in the stall flutter phenomenon:

$$\ddot{x} - a\dot{x} + \omega^2 x = f(x, \dot{x}), \quad x(0) = x_0, \quad \dot{x}(0) = x_1. \tag{106}$$

For sake of brevity in the explanations, we can assume without loss of generality that the function f is such that:

$$f(0, 0) = 0, \quad \frac{\partial f}{\partial \dot{x}}(0, 0) = 0, \quad \frac{\partial f}{\partial x}(0, 0) = 0. \tag{107}$$

7.1   Canonical Formulation for (98)

Let us set (it is assumed that $a < 2\omega$):

$$Y = \begin{pmatrix} x \\ \dot{x} \end{pmatrix}, \quad g(Y) = f(x, \dot{x}), \quad Y_0 = \begin{pmatrix} x_0 \\ x_1 \end{pmatrix}$$

and   (108)

$$A = \begin{pmatrix} 0 & 1 \\ -\omega^2 & a \end{pmatrix}, \quad F = \begin{pmatrix} 0 \\ g(Y) \end{pmatrix}.$$

Equation (111) is equivalent to the following one:

$$\frac{dY}{dt} = AY + F(Y), \quad Y(0) = Y_0. \tag{109}$$

The eigenvalues of A are complex and conjugate.

We set: $\lambda = \dfrac{a}{2} + i\omega\sqrt{1 - \dfrac{a^2}{4\omega^2}} = \alpha + i\beta$ and it is assumed that $0 < a << 2\omega$

in order to have amplified oscillations (which corresponds to a realistic case as for the Tacoma bridge treated at Sects. 2.6–2.9 or for the model of a military aircraft discussed in Sect. 8.

Let us now introduce a linear change of variables which enables one to diagonalize the matrix A. We set:

$$Y = PU, \quad U = \begin{pmatrix} u_1 \\ u_2 \end{pmatrix}, \quad P = \begin{pmatrix} 1 & 1 \\ \lambda & \bar{\lambda} \end{pmatrix}. \tag{110}$$

Equation (109) becomes:

$$\frac{dU}{dt} = \begin{pmatrix} \lambda & 0 \\ 0 & \bar{\lambda} \end{pmatrix} U + \frac{g(PU)}{\lambda - \bar{\lambda}} \begin{pmatrix} 1 \\ -1 \end{pmatrix}. \tag{111}$$

Therefore, the two relations contained in this vector equation are complex conjugate (hence $u_2 = \bar{u}_1$). Hence it is sufficient to treat one of these two. Let us choose the first one:

$$\frac{du_1}{dt} = \lambda u_1 + \frac{l(u_1, \bar{u}_1)}{\lambda - \bar{\lambda}}, \quad \text{with } l(u_1, \bar{u}_1) = g(PU). \tag{112}$$

7.2   Second Step: Elimination of Second Order Terms

Let us write the non linear term of Eq. (112) as follows:

$$l(u_1, \bar{u}_2) = k_{11}u_1^2 + k_{12}u_1\bar{u}_1 + k_{22}\bar{u}_1^2 + r(u_1, \bar{u}_1), \quad |r(u_1, \bar{u}_1)| = O(|u_1|^3).$$
(113)

A non linear change of variable is now used in order to cancel second order terms in the neighborhood of the origin. This leads to try the following expression *a priori*:

$$u_1 = z + p(z, \bar{z}), \quad z \in \mathbb{C}, \text{ and } p \in \mathscr{P}_2 \text{ homogeneous polynomials of degree 2.}$$
(114)

One has the following computational rules:

$$\begin{cases} \dfrac{du_1}{dt} = [1 + \dfrac{\partial p}{\partial z}(z, \bar{z})]\dot{z} + \dfrac{\partial p}{\partial \bar{z}}(z, \bar{z})\dot{\bar{z}} \\[4mm] \dfrac{d\bar{u}_1}{dt} = \dot{\bar{z}} + \ldots \text{ (at the first order).} \end{cases}$$
(115)

Hence:

$$\begin{cases} \dfrac{dz}{dt} = \lambda z + [\lambda p(z, \bar{z}) - \lambda\dfrac{\partial p}{\partial z}(z, \bar{z})z - \bar{\lambda}\dfrac{\partial p}{\partial \bar{z}}(z, \bar{z})\bar{z}] + \dfrac{k_{11}z^2 + k_{12}z\bar{z} + k_{22}\bar{z}^2}{\lambda - \bar{\lambda}} \\[4mm] \qquad\qquad +s(z, \bar{z}), \quad |s(z, \bar{z})| = O(|z|^3). \end{cases}$$
(116)

The question to be solved now is to find p so that all the second order terms could disappear. Let us introduce the linear endomorphism $\mathscr{L}$ of $\mathscr{P}_2$ defined by:

$$p(z, \bar{z}) = \alpha z^2 + \beta z\bar{z} + \gamma\bar{z}^2 \rightarrow \mathscr{L}(p) = \lambda p(z, \bar{z}) - \lambda\dfrac{\partial p}{\partial z}(z, \bar{z})z - \bar{\lambda}\dfrac{\partial p}{\partial \bar{z}}(z, \bar{z})\bar{z}.$$
(117)

Its matrix representation in the base $(z^2, z\bar{z}, \bar{z}^2)$ is the following one:

$$L_2 = \begin{pmatrix} -\lambda & 0 & 0 \\ 0 & -\bar{\lambda} & 0 \\ 0 & 0 & \lambda - 2\bar{\lambda} \end{pmatrix}.$$
(118)

Even for a small (and positive), this matrix is always invertible and maps a neighborhood of the origin (for u_1) into another one (for z). Hence the mapping remain local near the origin as far as the imaginary part of λ (the frequency) is non vanishing. Hence there exists $p \in \mathscr{P}_2$ such that one can suppress all the second order terms in Eq. (116). With this choice the system becomes:

$$\frac{dz}{dt} = \lambda z + s_3(z, \overline{z}). \tag{119}$$

7.3 Partial Elimination of Terms of Order 3

One can write:

$$s_3(z, \overline{z}) = k_{111}z^3 + k_{112}z^2\overline{z} + k_{122}z\overline{z}^2 + k_{222}\overline{z}^3 + j(z, \overline{z}), \quad \text{with } |j(z, \overline{z})| = O(|z|^4). \tag{120}$$

Following the same idea as in Sect. 7.2, we set ($\mathscr{P}_3$ is the space of third order homogeneous polynomials in z and $\overline{z}$):

$$z = \xi + q(\xi, \overline{\xi}), \quad \text{where } q \in \mathscr{P}_3 \tag{121}$$

This leads from a similar calculus as the one done in Sect. 7.2 to the following expression:

$$\begin{cases} \dfrac{d\xi}{dt} = \lambda\xi + [\lambda q(\xi, \overline{\xi}) - \lambda\dfrac{\partial q}{\partial \xi}(\xi, \overline{\xi})\xi - \overline{\lambda}\dfrac{\partial q}{\partial \overline{\xi}}(\xi, \overline{\xi})\overline{\xi}] \\[2mm] +k_{111}\xi^3 + k_{112}\xi^2\overline{\xi} + k_{122}\xi\overline{\xi}^2 + k_{222}\overline{\xi}^3 + j_4(\xi, \overline{\xi}) \\[2mm] \text{with } |j_4(\xi, \overline{\xi})| = O(|\xi|^4). \end{cases} \tag{122}$$

The matrix of the endomorphism:

$$q \in \mathscr{P}_3 \to \mathscr{L}(q) = \lambda q(\xi, \overline{\xi}) - \frac{\partial q}{\partial \xi}(\xi, \overline{\xi})\xi - \overline{\lambda}\frac{\partial q}{\partial \overline{\xi}}(\xi, \overline{\xi})\overline{\xi}, \tag{123}$$

in the basis $(\xi^3, \xi^2\overline{\xi}, \xi\overline{\xi}^2, \overline{\xi}^3)$ is the following one $(\lambda = \alpha + i\beta)$:

$$
L_3 = \begin{pmatrix}
-2\lambda & 0 & 0 & 0 \\
0 & -2\alpha & 0 & 0 \\
0 & 0 & -2\overline{\lambda} & 0 \\
0 & 0 & 0 & \lambda - 3\overline{\lambda}
\end{pmatrix}.
\tag{124}
$$

Because $a > 0$ can be small (beginning of an instability due to a negative damping) the matrix doesn't map a neighborhood of the origin (for z) into another neighborhood of the origin (for ξ). Therefore the term $\xi^2\overline{\xi}$ can't be eliminated locally. But the three other ones could be eliminated by this change of variable. For this reason the dynamical system becomes:

$$
\frac{d\xi}{dt} = \lambda\xi + h\xi^2\overline{\xi} + f_4(\xi, \overline{\xi}), \quad h = h_0 + ih_1, \quad \text{and} \ |f_4(\xi, \overline{\xi})| = O(|\xi|^4).
\tag{125}
$$

If there is a bounded solution remaining in a neighborhood of the origin of the so-called normal form:

$$
\frac{d\xi}{dt} = \lambda\xi + h\xi^2\overline{\xi}, \quad \xi(0) = \xi_0,
\tag{126}
$$

one can prove from a perturbation method and using a the first return mapping, that there is also one limit cycle of oscillations to (125) (see V.I. Arnold [1] and S. Wiggins [31]). But this can happen if and only if $h_0 < 0$.

Thus we are led to solve Eq. (126). Setting $\xi = \rho(t)e^{i\varphi(t)}$ in (126), one obtains:

$$
\begin{cases}
\dfrac{d\rho}{dt} = \alpha\rho + h_0\rho^3, \quad \rho(0) = |\xi(0)|, \\[3mm]
\dfrac{d\varphi}{dt} = \beta + h_1\rho^2, \quad \varphi(0) = \text{argument}(\xi(0)).
\end{cases}
\tag{127}
$$

We set (let us recall that $a = 2\alpha > 0$ and $h_0 < 0$):

$$
\rho_l = \sqrt{\frac{\alpha}{-h_0}}.
$$

The solution of the first equation (in ρ) is:

$$
\begin{cases}
\rho(t) = \dfrac{\rho(0)e^{\alpha t}}{\sqrt{1 + \dfrac{\rho(0)^2}{\rho_l^2}(e^{2\alpha t} - 1)}}, \\[2em]
\varphi(t) = \beta t - \displaystyle\int_0^t \rho(s)^2 ds \; t \geq 0.
\end{cases}
\tag{128}
$$

One has:

$$
\lim_{t\to\infty} \rho(t) = \rho_l, \quad \lim_{t\to\infty} \frac{d\varphi}{dt}(t) = \beta - \alpha\frac{h_1}{h_0}.
$$

Hence there is a limit cycle of oscillations defined by:

$$
\xi(t) = \rho_l e^{i(\beta - \alpha\frac{h_1}{h_0})t}.
\tag{129}
$$

Remark 19 In the complex plane the graph representation of the complex function $\xi(t)$ is a circle. The phase velocity is greater than β if $h_1 > 0$ and smaller if $h_1 < 0$. This splitting is due to the effect of the non linear terms. Let us point out again that this phenomenon only appears if $h_0 < 0$. If this is not the case, it is necessary to go on with the algorithm in order to find a resonant term.   □

Remark 20 If one would have $h_0 > 0$ instead of the case treated here, there is no third order limit cycle of oscillations. Hence it would be necessary to go further in the change of variables.

The elimination of all fourth order terms is straightforwards. The next so-called resonant terms would appear with terms of order five. And the analysis of these terms which can't be eliminated indicates if there is or not a limit cycle of oscillations at the order five. If this is not the case, the process should go on. In fact one can obtain a positive result only if there is a resonant term leading to a limit cycle of oscillations.   □

Remark 21 Assuming that one has a limit cycle of oscillations at the order three, the Fourier analysis of the signal $\xi(t)$ gives some informations concerning the spectrum of the solution $x(t)$. A resonance at the pulse $\omega_l = \beta - \alpha\frac{h_1}{h_0}$.

Because of the change of variable we did between z and ξ, the spectrum of z contains two resonant terms with the pulses ω_l and $3\omega_l$. The relation between u_1 and z enables one to claim that the spectrum of u_1 contains the resonant pulses ω_l, $2\omega_l$, $3\omega_l$, $4\omega_l$, $6\omega_l$.

Hence the fifth harmonic doesn't exist for such a limit cycle of oscillations. The diagram on Fig. 23 gives an illustration of this important characteristic which is a precious indicator of resonant terms of order three in a non linear and unstable

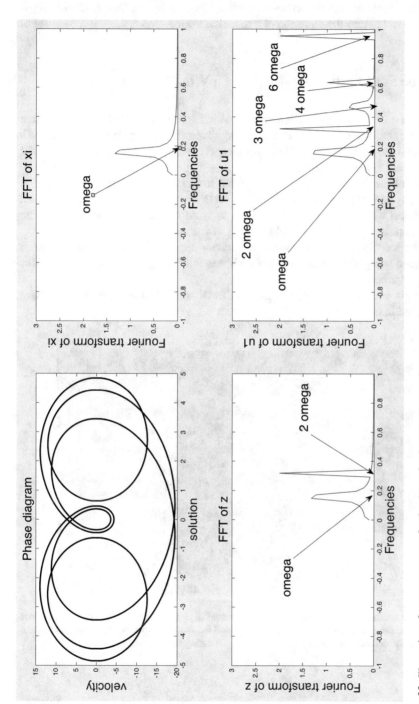

Fig. 23 Illustration of a resonant term of order three. On the right bottom (Fourier transform of the solution) one can see that the fifth harmonic doesn't exist. This statement is a characteristic of third order resonant terms. Obviously, if this term is not resonant ($h_0 \geq 0$) one should look for higher order resonant terms. Because all fourth order terms can be cancelled using a non linear change of variables, the first possibility would appear with the fifth order terms and so on

dynamical system. This property was point out initially by Ph. Destuynder and M.T. Ribereau in [17] for the vibration of a model in a wind tunnel.   □

8   The Stall Flutter of a Model in a Wind Tunnel

We discuss in this section a phenomenon which appeared in our wind tunnel when we performed some tests for an aircraft company. The model is at the scale 1/20. The tests have been performed only with respect to the angle of attack of the model with respect to the wind direction α.

The scheme of the model set in the wind tunnel is represented on Fig. 24. The fixation system of the model is performed using a scale shown on Fig. 25. The gust of each test was restricted to 30 s in order to avoid a fatigue breakdown due to the oscillations and the main flow velocity was $Mach = .7$ (Figs. 26, 27, 28, 29).

The recording of the aerodynamic forces applied by the model has been performed using a real time system which was reliable up to frequency 200 Hz for each component.

The six aerodynamic coefficients were measured but only c_x, c_z and c_{m0} have been used in our study.

Fig. 24 A military aircraft model in our wind tunnel at St Cyr l'Ecole. The angle of attack is driven from the rotation of the circular sector behind the model so that the center of mass which located at the center of the circle, remains unchanged during the rotation

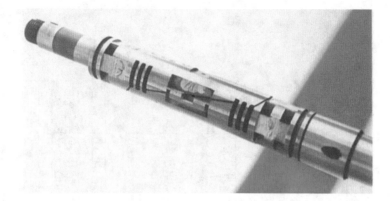

Fig. 25 The flexibility is due the six components scale which is weaken where the gages are set. One can see the transversal split on this picture. The rotation center of the scale system is slightly in the rear part of the model behind the center of mass of the model and the aerodynamic center

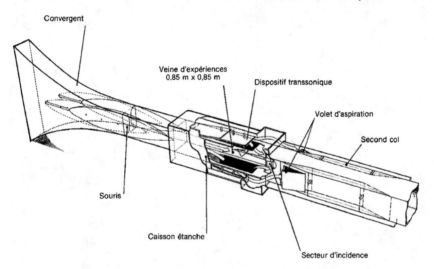

Fig. 26 The scheme of the wind tunnel used (Eiffel type with some improvements)

8.1   The Measures from the Wind Tunnel

Several quantities are measured versus the time.

- Measures of c_x, c_z, c_{m0}
- Pitching coefficient after smoothing by polynomials
- Position of the aerodynamic center (distance between O and P)

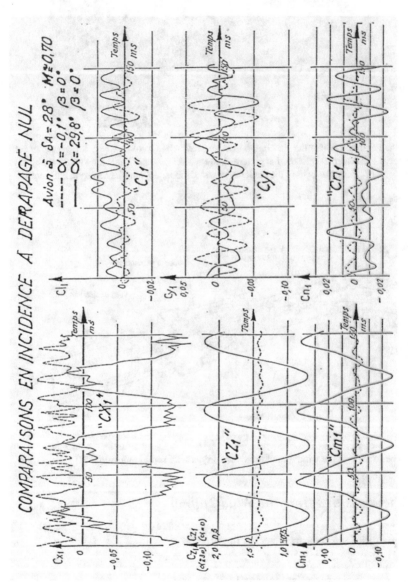

Fig. 27 The tests have been performed for two angles of attack: 29^0 which gives the largest values and the oscillations, and 0^0 for which only the drag coefficient is meaningful. One can see complex oscillations (more than one harmonic) on the coefficient c_{m0} which is connected to the bending moment at point O

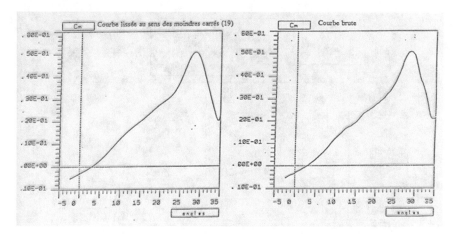

Fig. 28 On the left one can see the evolution of the pitching coefficient measured at point O but smoothed by splines functions and on the right the same curve without smoothing. The smoothing is necessary for the computation of the normal form at the order three

8.2 The Simple Mathematical Model Used

The pitching movement can be represented by the solution of the following equation:

$$J_0\ddot{\alpha} + C(\alpha - \alpha_0) = f(\alpha, \dot{\alpha}), \quad \text{with initial conditions.} \tag{130}$$

The right-hand side term f represents the moment of the aerodynamic forces at point O (see Fig. 30). The angle α_0 is the position at rest (without wind) of the model taking into account the moment of the weight. J_0 and C are respectively the moment of inertia of the model at point O and the stiffness of the supporting system (flexibility of the scale). The static equilibrium is referred by the angle α_e which is solution of:

$$C(\alpha_e - \alpha_0) = f(\alpha_e, 0). \tag{131}$$

The model can therefore be written around the equilibrium position $\alpha = \alpha_e$. Without confusion one can now formulate the aeroelastic model as follows:

$$\begin{cases} J_0\ddot{\alpha} + c(\alpha - \alpha_e) = f_e(\alpha, \dot{\alpha}), \quad \text{and initial conditions,} \\[2mm] \text{with } f_e(\alpha, \dot{\alpha}) = f(\alpha, \dot{\alpha}) - C(\alpha_e - \alpha_0), \quad \text{and } f_e(\alpha_e, 0) = 0. \end{cases} \tag{132}$$

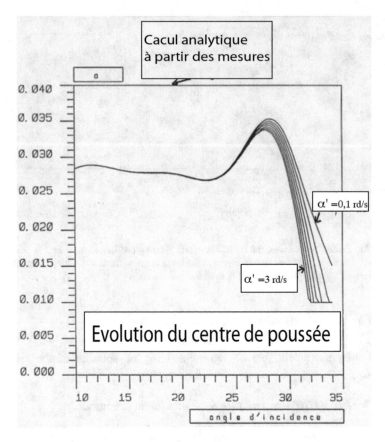

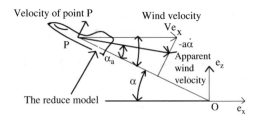

Fig. 29 The aerodynamic center is the point at which the the moment of the aerodynamic forces are zero (it exists in 2D). Because of the apparent angle of attack due to the apparent wind, this point is both dependent on the angle of attack but also its velocity. But it is almost constant for small angles of attack

Fig. 30 The mechanical system used to represent the pitching of the model

Let us first define the point P which is the center of aerodynamic forces. This point should satisfy the following relation, where α_a is the apparent angle of attack (see Fig. 30):

$$c_{m0}(\alpha_a) + a(c_x(\alpha_a)\sin(\alpha_a) - c_z(\alpha_a)\cos(\alpha_a)) = 0. \tag{133}$$

The value of a which is in this section the distance between O and P has been plotted on Fig. 29 for various value of $\dot{\alpha}$. One can see two things:

1. for angles of attack smaller than $25°$, the position of P is constant (experimentally speaking);
2. for angles larger than $28°$ the point P is going closer to the center of rotation O, which means in reality, that the forwards part of the main wing do not work very well concerning the lift. Therefore the center of aerodynamic forces is moving backwards on the aircraft. In fact this will imply a decrease of the moment c_{m0} (which can be observed on Fig. 28), and which can be interpreted as a drastic reduction of the so-called *Venturi effect* produced by the combination of the so-called *Canard-wing* and the main *Delta wing*.

The question is now to define $f(\alpha, \dot{\alpha})$ in order to take into account the apparent wind velocity. Let us first define the apparent wind at point P by (see Sect. 2.9):

$$v_a = V e_x - a\dot{\alpha}[\cos(\alpha)e_z + \sin(\alpha)e_x]. \tag{134}$$

The coefficient a corresponds to the point P defined previously. The modulus of the apparent velocity is:

$$\|v_a\| = V\sqrt{1 - 2a\sin(\alpha)\frac{\dot{\alpha}}{V} + a^2(\frac{\dot{\alpha}}{V})^2}, \tag{135}$$

and the apparent angle of attack is (see Sects. 2.6–2.9):

$$\alpha_a = \arctan(\tan(\alpha) - \frac{a\dot{\alpha}}{V\cos(\alpha)}). \tag{136}$$

Following the computations of Sects. 2.6–2.9, one derives from these formulae, the expression of $f(\alpha, \dot{\alpha})$ (S is the reference cross section and L a reference length both used in the definition of the aerodynamic coefficients):

$$f(\alpha, \dot{\alpha}) = \frac{\rho SL\|v_a\|^2}{2}c_{m0}(\alpha_a). \tag{137}$$

And a linearisation near $\alpha = \alpha_e$ of the model (131) leads to:

$$
\begin{cases}
J_0\ddot{\alpha}) + \dfrac{\rho a V S L}{2}\dot{\alpha}[2\sin(\alpha_e)c_{m0}(\alpha_e) + \cos(\alpha_e)\dfrac{\partial c_{m0}}{\partial\alpha}(\alpha_e)] + C(\alpha - \alpha_e) = 0, \\[12pt]
\text{with initial conditions.}
\end{cases}
$$

$$(138)$$

The so-called aerodynamic damping is positive (stability) if:

$$
\frac{\rho a V S L}{2}\dot{\alpha}[2\sin(\alpha_e)c_{m0}(\alpha_e) + \cos(\alpha_e)\frac{\partial c_{m0}}{\partial\alpha}(\alpha_e)] \geq 0. \qquad (139)
$$

The model is unstable if this quantity is strictly negative. Hence we have plot its evolution with respect to the angle of attack α on Fig. 31. A Fourier transform of the signals measured on the aerodynamic scale (on c_x and c_z for instance) plotted on Fig. 32 shows the gap for the fifth harmonic. This is fully in agreement with the theoretical analysis presented in Sect. 7. A numerical simulation based on

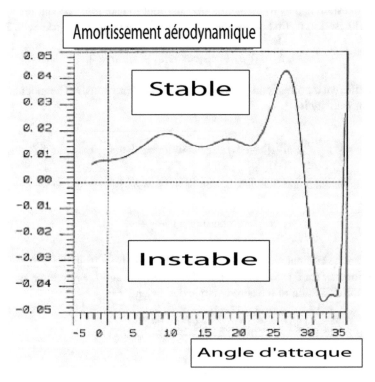

Fig. 31 Evolution of the aerodynamic damping with respect to the angle of attack. The damping is negative for $28° < \alpha < 35°$ which corresponds perfectly to what has been observed in the wind tunnel during the tests

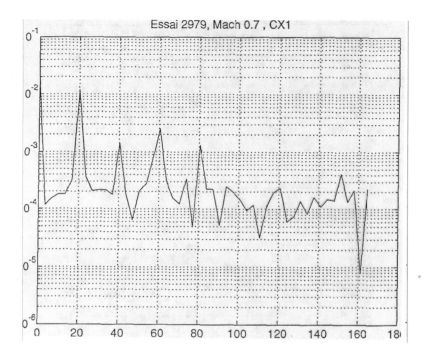

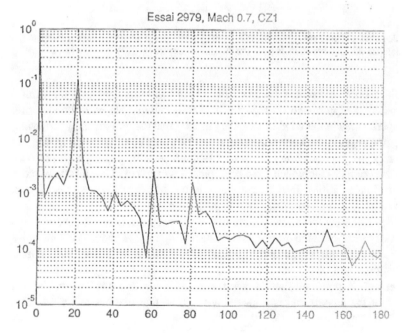

Fig. 32 For both graphs the FFT shows that there is no contribution on fifth harmonics (the fundamental frequency is here 20 Hz) but there are contributions on the second, the third, the fourth and the sixth as was predicted by the theory

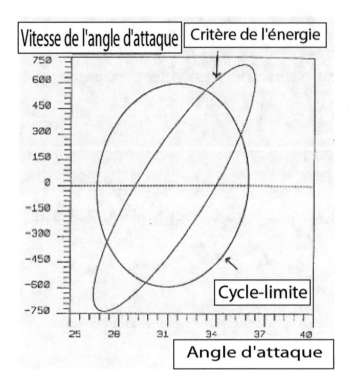

Fig. 33 On this figure are represented the limit cycle (computed) and the curve (computed) which separate the damped movements from the unstable ones (see Sect. 7)

the third order resonant term from Eq. (131) has been done and the limit cycle computed is represented on Fig. 33. It is fully compatible with the experimental tests performed in the wind tunnel. On this example the tool given by the dynamical system appears to be very accurate in order to predict what happens. It describes perfectly the phenomenon but also its magnitude with a quite high precision. In fact, the computation which have been presented in Sect. 7 are a little bit complex for hand calculations. In our case we used the software *Mathematica* in order to construct the normal form and finally, the limit cycle.

One important point which should be underlined here, is that it has been necessary to smooth the experimental data (c_x, c_z, c_{m0}) in order to obtain satisfying results in the construction of the normal form but even for the damping coefficient plotted on Fig. 31.

Let us also point out that the gap for the fifth harmonics is easy to check using a FFT of the signals and is a nice way to prove that one has really a third order resonant term.

9 About the Added Mass Method

The computation of quasi-static aerodynamic or hydrodynamic forces, requires to solve the Navier-Stokes equations. This is a very difficult and costly challenge which is often not very precise mainly due to instabilities in the flow. Another simplified approach consists in using the inviscid and incompressible framework. In this section we discuss briefly how this method enables one to take into account a so-called added mass to the structural model (Fig. 34).

In fact it is due to the kinematical energy communicated to the flow by the movements of the structure. Obviously, this is a coarse approach but it is much easier than solving Navier-Stokes equations. It enables one to introduce a velocity potential for the perturbation due to the movements of the structure -say φ- solution of the following model (where V_{e_x} is the velocity of the flow far from the structure and v is the unit normal along the boundary of Ω oriented outside of Ω which is the 2D open set occupied by the flow and contained in the plane $(O; e_1, e_2)$:

$$
\begin{cases}
-\Delta\varphi = 0 \text{ in the open set } \Omega \text{ surrounding the structure,} \\[2mm]
\dfrac{\partial\varphi}{\partial v} = u.v \text{ on the boundary } \Gamma_S \text{ of the structure } (u \text{ is the velocity}), \\[2mm]
u \text{ is a function of time,} \\[2mm]
\varphi = 0 \text{ on } \Gamma_0 \text{ where a pressure is prescribed,} \\[2mm]
\dfrac{\partial\varphi}{\partial v} = V(v.e_x) \text{ on the exterior boundary of } \Omega.
\end{cases}
\tag{140}
$$

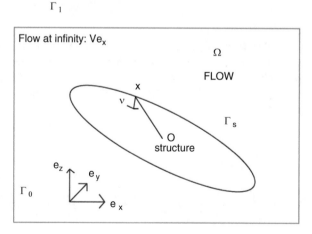

Fig. 34 The structure immersed in a flow (2D in our case) prescribes a displacement of this flow when it is moving. This increases the inertia of the structure in the model discussed in this Sect. 9

The solution φ of (140) is uniquely defined as a linear expression of $u.v$ as soon as $\overset{\circ}{\Gamma}_0$ is not empty and u (normal velocity of the boundary of the structure) smooth enough. Let us introduce φ_0 solution of (140) if $u = 0$ and $V = 1$. Hence one can set:

$$\varphi = G(u.v) + V\varphi_0. \tag{141}$$

Because the linear operator G is time-independent, one has:

$$\frac{\partial \varphi}{\partial t} = G(\frac{\partial u}{\partial t}.v). \tag{142}$$

In fact, the velocity u is the time derivative of the displacement denoted by d of the structure. Hence, assuming small enough perturbations in order to neglect convection terms, one has:

$$\frac{\partial u}{\partial t} = \frac{\partial^2 d}{\partial t^2}, \tag{143}$$

The modifications of the pressure applied on the structure due to its movement in the framework of small perturbations, are (ρ is assumed to be constant in this formulation):

$$p = -\rho \frac{\partial \varphi}{\partial t} = -\rho G(\frac{\partial^2 d}{\partial t^2}.v). \tag{144}$$

Assuming again that the displacements of the structure are small one can use a kinematical torsor representation by setting ($\wedge$ is the vector product and $(., ., .)$ is the mixed vector product):

$$d = d_0 + \Theta \wedge Ox, \tag{145}$$

where d_0 is the displacement of point O and Θ is the rotation vector (only valid if the displacements are small). The resultant transient force torsor at point O applied to the structure by the flow and due to its velocity implying a modification of the pressure in the flow, are at point O (e_3) is the unit vector normal to the plane (e_1, e_2) parallel to Θ and Ψ:

$$\begin{cases} F_{ad} = -\rho \int_{\Gamma_S} [G(\frac{\partial^2 d_0}{\partial t^2}.v) + G(\frac{\partial^2 \Theta}{\partial t^2}, Ox, v))]v, \\[3mm] M_{ad}(O) = -\rho \int_{\Gamma_S} [G(\frac{\partial^2 d_0}{\partial t^2}.v) + G((\frac{\partial^2 \Theta}{\partial t^2}, Ox, v))](Ox, v, e_3). \end{cases} \tag{146}$$

For sake of brevity we consider that the point O is the center of mass G of the rigid structure. Let M be the mass of the structure and J_G the inertia tensor at G. Let us assume that the structure is fixed through a system of springs (tension and torsion) denoted respectively by

$$\mathcal{K} = \begin{pmatrix} A_G & 0 \\ 0 & C_G \end{pmatrix} \quad \text{and} \quad \mathcal{M} = \begin{pmatrix} M & 0 \\ 0 & J_G \end{pmatrix}.$$

The external forces applied to the system (if there are) are denoted by

$$\mathcal{F}_{ad} = \begin{pmatrix} F_{ext} \\ M_{ext} \end{pmatrix},$$

the dynamic aerodynamic torsor (which depends on time). It can depend on d_G and Θ because of the aerodynamic forces (see previous sections). We also introduce the notation

$$X = \begin{pmatrix} d_G \\ \Theta \end{pmatrix} \quad \text{and} \quad Y = \begin{pmatrix} \delta_G \\ \psi \end{pmatrix}. \tag{147}$$

The equations of the movement are the following ones:

$$\begin{cases} M\dfrac{\partial^2 d_G}{\partial t^2} + A_G d_G = F_{ext} + F_{ad} \\[2mm] = F_{ext} - \rho \displaystyle\int_{\Gamma_S} [G(\dfrac{\partial^2 d_G}{\partial t^2}.v) + G(\dfrac{\partial^2 \Theta}{\partial t^2}, Ox, v)]v, \\[4mm] J_G\dfrac{\partial^2 \Theta}{\partial t^2} + C_G\Theta = M_{ext} + M_{ad}(O) \\[2mm] = M_{ext} - \rho \displaystyle\int_{\Gamma_S} [G(\dfrac{\partial^2 d_G}{\partial t^2}.v) + G((\dfrac{\partial^2 \Theta}{\partial t^2}, Ox, v))](Ox \wedge v). \end{cases} \tag{148}$$

We set for any rigid body displacement of the structure $\delta = \delta_G + \Psi \wedge Ox$ (be careful about the change of sign and let us notice that the two first following vectors are orthogonal to the two last ones):

$$\begin{cases} \mathcal{G}_{ad}(Y) = \rho \displaystyle\int_{\Gamma_S} G(\delta_G.v)v, \quad \mathcal{C}_{ad}(Y) = \rho \displaystyle\int_{\Gamma_S} G((\psi, Ox, v)v), \\[4mm] \mathcal{J}_{ad}(Y) = \rho \displaystyle\int_{\Gamma_S} G((\Psi, Ox, v))(Ox \wedge v), \quad \mathcal{D}_{ad}(Y) = \rho \displaystyle\int_{\Gamma_S} G(\delta_G.v)(Ox \wedge v). \end{cases} \tag{149}$$

We introduce the following bilinear form:

$$\mathcal{Q}_{ad}(X, Y) = \mathcal{G}_{ad}(X).\delta_G + \mathcal{C}_{ad}(X).\delta_G + \mathcal{J}_{ad}(X).\psi + \mathcal{D}_{ad}(X).\psi, \qquad (150)$$

and the linear operator linked to this bilinear form, called the added mass:

$$\mathcal{M}_{ad}(X).Y = \mathcal{Q}_{ad}(X, Y) \qquad (151)$$

The equations of the movement of the structure developed in the previous sections, can therefore be modified as follows:

$$(\mathcal{M} + \mathcal{M}_{ad})(\frac{\partial^2 X}{\partial t^2}) + \mathcal{K}X = \mathcal{F}_{ext} = \begin{pmatrix} F_{ext} \\ M_{ext} \end{pmatrix}. \qquad (152)$$

Remark 22 The static pressure due to the term $V\varphi_0$ is already taken into account in the expression of the aerodynamic forces applied to the structure. It is therefore not included in the term on the right-hand side of Eq. (152) which only contains transient terms. But one could suggest to upgrade the contribution of φ_0 by using more accurate expressions for the aerodynamic forces. The model given in this section is only valid for the transient analysis and it doesn't take into account the dependence of the aerodynamic stiffness and the aerodynamic damping (see previous sections).   □

The new term appearing in the inertia is called the added mass. The following Theorem 5, justifies the terminology *added*. In fact, the movement of the structure implies a displacement of the flow surrounding and therefore it acts as if the mass was increased (more or less depending on the movement).

Theorem 5 *Let $d = d_G + \Theta \wedge Ox$ and $\delta = \delta_G + \psi \wedge Ox$ be two displacement fields of the structure (rigid body motion in small displacement). Let X and Y be the vectors of $\mathbb{R}^6$ associate to these displacements as defined at (147) and $\mathcal{Q}_{ad}$ the bilinear form defined at (150). One has:*

$$\mathcal{Q}_{ad}(X, Y) = \mathcal{Q}_{ad}(Y, X), \qquad (153)$$

and

$$\mathcal{Q}_{ad}(X, X) \geq 0. \qquad (154)$$

□

Proof Let $d = d_G + \Theta \wedge Ox$ and $\delta = \delta_G + \psi \wedge Ox$ be two displacement fields of the structure One has:

$$\mathcal{Q}_{ad}(X, Y) = \rho \int_{\Gamma_S} G(d_G.v + (\Theta, Ox, v))(\delta_G.v + (\Psi, Ox, v)). \tag{155}$$

Let us now introduce the two functions φ_1 and φ_2 solutions of:

$$\begin{cases} -\Delta\varphi_1 = 0 \text{ in } \Omega, \\[2mm] \dfrac{\partial\varphi_1}{\partial v} = 0 \text{ on } \Gamma_1, \ \varphi_1 = 0 \text{ on } \Gamma_0, \\[2mm] \dfrac{\partial\varphi_1}{\partial v} = (d_G + \Theta \wedge Ox).v \text{ on } \Gamma_S, \end{cases} \qquad \begin{cases} -\Delta\varphi_2 = 0 \text{ in } \Omega, \\[2mm] \dfrac{\partial\varphi_2}{\partial v} = 0 \text{ on } \Gamma_1, \ \varphi_2 = 0 \text{ on } \Gamma_0, \\[2mm] \dfrac{\partial\varphi_2}{\partial v} = (\delta_G + \Psi \wedge Ox)).v \text{ on } \Gamma_S, \end{cases}$$
$$\tag{156}$$

One has the following relation from a classical integration by parts:

$$\mathcal{Q}_{ad}(X, Y) = \int_\Omega \nabla\varphi_1.\nabla\varphi_2. \tag{157}$$

It proves that (symmetry) $\mathcal{Q}_{ad}(X, Y) = \mathcal{Q}_{ad}(Y, X)$ and (positivity and even $H^1(\Omega)$ coercivenes):

$$\mathcal{Q}_{ad}(X, X) \geq 0.$$

This why one can say that the new term appearing in the inertia of the structure is an added mass. □

Remark 23 The strategy that we have presented here is restricted to the case of an infinitesimal rigid body motion for sake of brevity and because it is an important case in the applications. But it can be extended without difficulty to flexible structures and even with large displacements. □

10 Conclusion

This text was devoted to a brief survey of some aeroelastic problems in the framework of the quasi-static approximation. Our main subject was the study of instabilities due to aerodynamic negative damping. This phenomenon appears when a stall flutter occurs and it is strongly connected to the apparent wind velocity. If the frequencies of the structure involved are sufficiently low (small reduce frequencies), one can make use of a large number of mathematical tools which enable to describe precisely what happens. In particular the possibility of limit cycle of oscillations. Let us mention a corner stone of these tools; it is the normal form method which is very accurate and reliable. Furthermore, the vibrations observed on flexible mechanical systems can be reduced and even suppressed using a control strategy. Here again the existing tools are very efficient. A short description is given using optimal control theory which can degenerate into an exact control if some controllability hypotheses are satisfied. In some cases, the non linear vibrations (limit cycle of oscillations) are localized in a part of the full structure immersed in a flow. Furthermore, the frequencies can be very different from one substructure to the other. The possibility of using Nash equilibrium points is a very stimulating possibility in order to decentralize the controls. In this text we choose to focus on decentralized exact control with coordination which is easier to handle. All these aspects have been discussed in the present text and a coordination algorithm is suggested. In the whole text, we used two industrial examples as supports for the understanding of our theory, in order to illustrate the theoretical developments. One is the famous Tacoma Narrows bridge for which so many very different explanations have been written and the second one, is a model of a military aircraft that we have tested in our aerodynamic wind tunnel in Saint-Cyr l'Ecole and which strangely, behaved very similarly to the Tacoma Narrows bridge.

References

1. Arnold, V.: Dynamical System, nine volumes (under the supervision of V. Arnold). Encyclopaedia of Mathematical Sciences Series. Springer, New York (1990)
2. Balakrishnan, A.V.: Aeroelasticity: The Continuum Theory. Springer, New York (2012)
3. Bellman, R.: Control Theory. Sci. Am. 211(3), 186–200 (1964)
4. Bellman, R.E.: Dynamic Programming. Princeton University Press, Princeton, NJ (1957)
5. Bellman, R.E.: Adaptive Control Processes: A Guided Tour. Princeton University Press, Princeton, NJ (1961)
6. Bensoussan, A.: Points de Nash dans le cas de fonctionnelles quadratiques et jeux di fférentiels linéaires à N personnes. SIAM J. Control 12, 3 (1974)
7. Bensoussan, A., Lions, J.L.: Application of Variational Inequalities in Stochastic Control. North Holland, Amsterdam (1982)
8. Billah, Y.K., Scanlan, R.H.: Resonance, Tacoma bridge failure, and undergraduate physics textbooks. Am. J. Phys. 59(2), 118–124 (1991)
9. Bisplinghoff, R.L., Ashley, H., Halfman, H.: Aeroelasticity. Dover Science, New York (1996)
10. Cartan, E.: Calcul différentiel. Herman, Paris (1957)
11. Cea, J.: Optimisation: Théorie et algorithmes. Dunod, Paris (1968)
12. Collar, A.R.: The first fifty years of aeroelasticity. Aerospace 5(2), 12–20 (1978)
13. Curtiss, H.C.Jr., Scanlan, R.H., Sisto, F., Dowell, E.H.: A Modern Course in Aeroelasticity. (Mechanics: Dynamical Systems), 2nd rev. and enlarged Edition. Dover, New York (2008)
14. Den Hartog, J.P.: Mechanical Vibrations, 3rd edn. McGraw-Hill, New York (1947)
15. Destuynder, Ph.: Analyse et contrôle des équations différentielles ordinaires. Hermès-Lavoisier, Londres-Paris (2006)
16. Destuynder, Ph.: Aéroélasticité et aéroacoustique. Paperback Hermès-Lavoisier, Londres-Paris (2007)
17. Destuynder, Ph., Ribereau, M.T.: Non linear dynamics of test models in wind tunnels. Eur. J. Mech. A/Solids 15(1), 91–136 (1996)

18. Fung, Y.C.: An introduction to the theory of aeroelasticity. In: Dover Books on Aeronautical Engineering. Dover, New York (1969)
19. Garrick, I.E., Reed, W.H.: Historical development of aircraft flutter. J. Aircraft **18**, 897–912 (1981)
20. Glowinski, R., Lions, J.L., Trémolières, R.: Analyse numérique des inéquations variationnelles. Tomes I et II, Dunod, Paris (1976)
21. Hoque, M.E.: Active Flutter Control. LAP Lambert Academic Publishing, Saarbrücken (2010)
22. Kalman, R.E.: Mathematical description of linear dynamical systems. J. Soc. Ind. Appl. Math. **1**, 152–192 (1963)
23. Lions, J.L.: Contrôle optimal des systèmes gouvernés par des équations aux dérivées partielles. Dunod, Paris (1968)
24. Lions, J.L.: Contrôlabilité exacte, perturbations et stabilisation de systèmes distribués. Masson, Paris (1988)
25. Pallu de la Barrière, R.: Cours d'automatique théorique. Dunod, Paris (1966)
26. Pontryagin, L.S.: Equations différentielles ordinaires. Editions Mir, Moscou (1969)
27. Pontryagin, L.S., Boltyanskii, V.G., Gamkrelidze, R.V., Mishchenko, E.F.: The Mathematical Theory of Optimal Processes. Interscience and ed. Mir, Moscou (1962)
28. Roseau, M.: Vibrations des systèmes mécaniques: méthodes analytiques et applications. Masson, Paris (1984)
29. Scanlan, R.H.: Wind Effects on Structures: An Introduction to Wind Engineering, 2nd edn. Wiley-Interscience, New York (1986)
30. Tikhonov, A., Arsenin, V.: Solution of Ill-Posed Problems. Winston and Sons, Washington (1977). ISBN 0-470-99124-0
31. Wiggins, S.: Introduction to Applied Nonlinear Dynamical Systems and Chaos. Springer, Berlin (1990)
32. Wright, J.R., Cooper, J.E.: Introduction to Aircraft Aeroelasticity and Loads. Wiley, New York (2007)

Computational Treatment of Interface Dynamics via Phase-Field Modeling

Miguel Bures, Adrian Moure, and Hector Gomez

Abstract This chapter describes the phase-field approach to modeling interface dynamics, with particular emphasis on the mathematical formulation and computational aspects. We describe to approaches to derive a phase-field model. The first one can be understood as a regularization approach: We start with a sharp interface model which is later replaced by a diffuse interface. The second approach starts with a free energy functional and balance laws for mass, linear momentum and energy; the final governing equations are derived using the second law of thermodynamics and the Coleman-Noll procedure. We finish by illustrating how the phase-field method can be used to solve problems on complicated geometries using cartesian grids only. Some of the opportunities opened by the phase-field approach are illustrated with numerical simulations.

Keywords Interfacial mechanics · Phase-field modeling · Isogeometric analysis

1 Introduction

1.1 Interface Dynamics in Computational Mechanics and the Phase-Field Method

Moving boundary problems are common in several research areas and have been addressed from different approaches. However, the efficient solution of moving boundary problems is still a challenge. From a computational point of view, these problems often comprise a set of partial-differential equations (PDEs) posed on moving domains. The set of PDEs is coupled with another set of boundary conditions which are posed on a moving interface whose location is an unknown

M. Bures · A. Moure · H. Gomez (✉)
School of Mechanical Engineering, Purdue University, West Lafayette, IN, USA
e-mail: mburesmu@purdue.edu; amourero@purdue.edu; hectorgomez@purdue.edu

© The Author(s), under exclusive license to Springer Nature Switzerland AG 2021
D. Greiner et al. (eds.), *Numerical Simulation in Physics and Engineering: Trends and Applications*, SEMA SIMAI Springer Series 24,
https://doi.org/10.1007/978-3-030-62543-6_2

of the problem. This fact significantly increases the difficulty of the problem. Some examples of moving boundary problems are crack propagation and fluid-structure interaction. Though some of these problems have been successfully addressed, many others are yet to be solved efficiently.

A new approach to moving boundary problems which has recently been introduced is the phase-field method [3, 13, 20, 22, 54, 55, 59]. The phase-field method relies on the reformulation of the original moving boundary problem such that the PDEs are posed on a fixed computational domain rather than on a moving domain. This method avoids the use of complex moving domains and the explicit tracking of the interface. To accomplish this, we have to introduce a new variable, named *phase field*, defined in the whole computational domain, which serves to determine at each point the phase that is present.[1] The main difference between the phase-field approach and the moving boundary problem is that the former has a small but finite interface that separates the different phases. For this reason, the phase-field approach is also known as the *diffuse-interface* model while the original moving boundary problem is known as the *sharp-interface* model. Besides avoiding moving domains, the phase-field method permits us to obtain smooth interfaces and hence, smooth solutions compared to the sharp-interface method. In some cases, the phase-field method converges to the moving boundary problem as the parameter that regulates the interface width tends to zero. There are several ways to derive the diffuse-interface problem equivalent to a moving boundary problem. One option consists in smoothing out (or *diffusifying*) the corresponding moving boundary problem. This approach will be explained in Sect. 2. After describing the *diffusification* of a solidification model, we will indicate how to prove mathematically its convergence to the sharp-interface problem using the *matched asymptotic expansions* theory [11, 24]. Although this may be regarded as the simplest way to obtain a phase-field model from a moving boundary problem, there are other methods to do so. In fact, phase-field models do not require a moving boundary problem associated with them. A completely different approach consists in deriving the model directly from free energy functionals using a thermomechanical approach and the Coleman-Noll procedure [15, 63]. In Sect. 3, we will show how some basic phase-field models can be derived from a thermomechanical point of view. In particular, we will derive the classical Allen-Cahn and Cahn-Hilliard models and the Navier-Stokes-Cahn-Hilliard model for two-component fluid flow.

One of the key advantages of the phase-field method is that the tracking of the interface is not necessary, which makes the problem easier and computationally more efficient. Nonetheless, the phase-field method introduces a new variable and, consequently, a new PDE, which entails other drawbacks that need to be tackled. The first problem stems from the nature of the phase-field equation. The phase-field equation may have fourth-(or even higher) order derivatives. With simple

---

[1]The term phase may refer to different concepts. It can be used, e.g., to define the different states of matter in a solid-liquid model, the different components in a mixture, or to determine whether there is a fracture or not in a crack propagation model.

geometries, there are several numerical methods, for instance the finite difference method, which allow us to solve such problems. In the case of complex geometries, those numerical schemes cannot be easily employed. In that case, the finite element method is the usual choice. However, the finite element method presents limitations when applied to higher-order PDEs. Classical finite element methods require basis functions of, at least, continuity C^{m-1} to solve a PDE with highest derivatives of order $2m$. Recently, a new methodology called isogeometric analysis [17, 38, 41, 42] has been introduced. Isogeometric analysis allows us to create basis functions with adaptable continuity even in complex geometries. Another challenge is the treatment of the diffuse interface and the spatial approximation of the solution in that region. Since we usually resort to thin interfaces, the phase field transitions quickly from one phase to another in the surroundings of the interface, producing large phase field gradients. This quick transition must be captured by the computational mesh, which must have a minimum number of elements in the interface.

Currently, thanks to the computational improvements in the past years and the theoretical development of the phase-field methodology, phase-field modeling can be regarded as a powerful tool to predict the behavior of a great number of problems that where challenging or extremely difficult to solve using traditional methodologies. Just to mention a few, the most relevant problems include but are not limited to crack propagation and crack interaction in complex three-dimensional geometries [6], three-dimensional air-water flow with surface tension [12], liquid-vapor phase transitions and cavitation [50], nucleate and film boiling [51], phase-change-driven implosion of thin structures [10], cellular migration [62], and tumor growth [39, 52, 64–66, 68].

The contents of this chapter are organized as follows. After this introduction we include a brief subsection to clarify notational conventions. In Sect. 2, we treat the derivation of phase-field models from moving boundary problems in the context of solidification, using the process of diffusification. Then, in Sect. 3 we consider the derivation of phase-field models from a thermomechanical perspective and derive a variety of models that are common in computational mechanics. In Sect. 4, we apply the diffuse domain approach, a technique to solve PDEs on moving domains using a fixed underlying mesh, to transform a convection-diffusion problem into a phase-field model. Finally, in Sect. 5, we present some numerical examples.

1.2 Notational Conventions

During this chapter, we will define certain functions, for instance f, on a real interval I. This function transforms any x belonging to I into $f(x)$. Its spatial derivative will be denoted as f' or $\frac{df}{dx}$. We may also define Ω as a subset of $\mathbb{R}^{n_d}$, where n_d is the number of spatial dimensions. The boundary of the subset will be denoted as $\partial\Omega$ or Γ. The variable x denotes a spatial point, with $x = (x_1, x_2, \ldots, x_{n_d})$. The function u denotes a velocity field. Throughout this chapter, the variable t will be used to express time. Functions often depend on more than one variable, for instance, $\phi(x, t)$. In that case, we denote their partial derivatives as $\partial_i \phi = \frac{\partial \phi}{\partial x_i}$, $\partial_t \phi = \frac{\partial \phi}{\partial t}$.

We may also use the operator ∇, meaning $\nabla \phi$ the spatial gradient of ϕ. On the other hand, the operator $\text{div}(\cdot)$ defines the divergence of a field, for instance, $\text{div } u$ denotes the divergence of the velocity field u. The material derivative will be also used in this chapter. The material derivative of a function ϕ with a field velocity u may be expressed as $\dot{\phi} = \partial_t \phi + u \cdot \nabla \phi$. Finally, we will define the variational derivative of a functional $\mathcal{F}(\phi, \nabla \phi) = \int_\Omega \Phi(\phi, \nabla \phi) \, dx$ as

$$\frac{\delta \mathcal{F}}{\delta \phi} := \frac{\partial}{\partial \phi} \Phi(\phi, \nabla \phi) - \text{div} \left(\frac{\partial}{\partial (\nabla \phi)} \Phi(\phi, \nabla \phi) \right).$$

It is important to mention that, throughout this chapter, some functions, variables, or tensors are used in multiple sections, e.g., the Helmholtz free energy or the double well potential. Although the name and denomination remains the same, the reader should not regard them as the same quantity for all sections. To avoid confusion, these terms are described in each section.

2 Derivation of Phase-Field Models via Diffusification

The goal of this section is to show how we can *diffusify* classical moving boundary problems to obtain the equivalent phase-field models. The phase-field model displays a small but finite interface and is expressed as a set of PDEs posed on a fixed domain. With the *diffusification* procedure, the sharp interface of the moving boundary problem is transformed into a diffuse interface.

To introduce the concept of *diffusification*, we will derive a phase-field model for a classical solidification problem. We consider a solidification theory for pure materials, known as the *generalized Stefan problem*. The original moving boundary problem consists of two PDEs posed on different domains, corresponding to solid and liquid phases. The phases are adjacent and separated by a sharp interface that moves with time. The location of the interface is one of the unknowns of the process. To track the interface motion, both PDEs have to be coupled with appropriate boundary conditions held on the interface. We will now proceed to *diffusify* this problem to obtain the equivalent phase-field model composed of two PDEs posed on a fixed domain.

2.1 A Classical Moving Boundary Problem for Solidification of Pure Materials

We consider a system composed of a single material that presents two different phases, i.e., the solid and liquid phases. The material can undergo phase transformations between solid and liquid. Let us consider the solid-liquid system located in a fixed spatial domain Ω. The domain can be decomposed in two subdomains such that $\overline{\Omega} = \overline{\Omega_S \cup \Omega_L}$, where Ω_S is the region of Ω occupied by the solid phase, and Ω_L the region occupied by the liquid phase. We have to determine the interface between both phases, named Γ_{LS}, located where Ω_L and Ω_S meet. Although Ω is fixed, note the two subsets Ω_S and Ω_L may change and, thus, the motion of the interface is also an unknown of the problem. Consequently, the moving boundary problem is written as

$$\frac{\partial e}{\partial t} + \text{div}\, \boldsymbol{q} = 0 \text{ in } \Omega_S \cup \Omega_L, \tag{1}$$

$$\lambda V_n = [\![\boldsymbol{q}]\!] \cdot \boldsymbol{n}_{LS} \text{ on } \Gamma_{LS}, \tag{2}$$

$$[\![\theta]\!] = 0 \text{ on } \Gamma_{LS}, \tag{3}$$

$$\sigma(\omega V_n + \overline{\kappa}) = H\frac{\theta_m - \theta}{\theta_m} \text{ on } \Gamma_{LS}. \tag{4}$$

Note that in this formulation, for simplicity, we omitted the boundary conditions on $\partial\Omega$. The problem defined by Eqs. (1)–(4) is known as the generalized Stefan problem.[2] Equations (1) and (2) simply represent an energy balance on a body with discontinuous properties. Equation (3) states that temperature is continuous across the interface. Equation (4) represents the *Gibbs-Thomson condition*, which is used to determine the motion of the interface. Equation (1) is a classical energy balance where e is the internal energy and $\boldsymbol{q}$ is the heat flux. These two variables are not continuous across the interface due to the change in properties between the two phases. We can define the internal energy as

$$e = C_v\theta + \lambda\chi_L, \tag{5}$$

where C_v is the specific heat capacity per unit of volume, θ the temperature, and λ the latent heat per unit of volume. χ_L can be regarded as a marker of the liquid phase, i.e., a function defined in the whole domain Ω, which takes the value 1 in Ω_L and 0 in Ω_S. To define the heat flux we use Fourier's law,[3] i.e., $\boldsymbol{q} = -k\nabla\theta$. Here, the conductivity k can take different values in Ω_S and Ω_L, denoted as k_S and k_L, respectively. In Eq. (2), $\boldsymbol{n}_{LS}$ is the unit normal to Γ_{LS}, defined as positive when pointing towards the liquid. V_n is the velocity of the interface in the direction of $\boldsymbol{n}_{LS}$ and $[\![\cdot]\!]$ denotes the jump across the interface Γ_{LS}, i.e., $[\![\boldsymbol{q}]\!] = \boldsymbol{q}_S - \boldsymbol{q}_L$. Finally, in Eq. (4), ω is the coefficient of kinetic undercooling, σ is the surface tension of the interface, θ_m is the melting temperature, H represents the interfacial enthalpy per

---

[2] In the case $\sigma = 0$, Eqs. (1)–(4) is referred to as the *classical Stefan problem*.
[3] Although Fourier's law is the most common theory used in heat transmission problems, there are other possibilities [28, 29, 31, 32].

unit volume, and $\bar{\kappa}$ is the sum of principal curvatures of Γ_{LS}. In this chapter, we assume negative values of $\bar{\kappa}$ for spherical liquid phase droplets.

2.2 The Basis of Diffusification

So far, we have defined a PDE in Eq. (1), posed in the two domains Ω_L and Ω_S, and a set of boundary conditions, Eqs. (2)–(4), defined on the interface between both subdomains. We can interpret the role of Eqs. (2)–(4) as one boundary condition for the PDE on Ω_S, another for the PDE on Ω_L, and the last one to determine the motion of the interface Γ_{LS}. It is also interesting that in the particular case $\sigma = 0$, i.e., the classical Stefan problem, Eq. (4) is simplified to $\theta = \theta_m$. This means that the temperature in the interface is the melting temperature. In that case, the interface is automatically identified by computing the temperature field. In the general case, however, the location of the interface cannot be inferred from temperature only, and further information is needed. That information will be encoded in the *phase field*, which will act as a marker of the position of the different phases and the interface.

To define the phase field we proceed to establish a set of constraints. The phase field should acquire constant values at the phases, but different on each one. In addition, it should have a small but smooth transition across the interface between the two phases. Thus, our phase-field function, named ϕ, must be defined in the entire domain Ω and depend on spatial position and time, i.e., $\phi = \phi(x, t)$. Commonly, the convention is that ϕ takes the value 1 at one phase and -1 at the other. Another common option is using $\phi = 0$ and $\phi = 1$ to define the pure phases. In our case, we will use the value 1 for the solid phase and -1 for the liquid phase. As an assumption, we will define ϕ as a hyperbolic tangent

$$\phi(x, t) = f\left(\frac{d_t(x)}{\sqrt{2}\epsilon}\right) := \tanh\left(\frac{d_t(x)}{\sqrt{2}\epsilon}\right). \tag{6}$$

Here, $d_t(x)$ represents the signed distance between point x and the interface Γ_{LS}, with the convention of positive sign if x is in the solid phase and negative if it is in the liquid phase. The parameter ϵ controls the interface width. Note that in the equivalent moving boundary problem, the interface would be defined by the set of points x such that $\phi(x, t) = 0$. From Eq. (6), we will derive a phase-field model where the phase field naturally converges to the hyperbolic tangent profile as $\epsilon \to 0$. Latter, we will prove this fact.

2.3 The Phase-Field Model for Solidification

To diffusify the generalized Stefan problem we start with Eq. (4), but first we will define some quantities and functions. We define $\bar{\kappa}$ in terms of the phase field. For simplicity, we remove the subscript t from d_t in Eq. (6). It is known that $|\nabla d| = 1$, $\mathbf{n}_{LS} = \nabla d$ at Γ_{LS}, and the curvature κ satisfies $\kappa = \nabla\nabla d$ at Γ_{LS}; see, e.g., [19]. From Eq. (6), we can compute the spatial derivatives of ϕ as

$$\partial_i\phi = \frac{1}{\sqrt{2}\epsilon} f'\left(\frac{d}{\sqrt{2}\epsilon}\right)\partial_i d, \tag{7}$$

$$\partial_{ij}^2\phi = \frac{1}{2\epsilon^2} f''\left(\frac{d}{\sqrt{2}\epsilon}\right)\partial_i d\partial_j d + \frac{1}{\sqrt{2}\epsilon} f'\left(\frac{d}{\sqrt{2}\epsilon}\right)\partial_{ij}^2 d. \tag{8}$$

It is known that, if f is a hyperbolic tangent function, then $f' = 1 - f^2$ and $f'' = -2ff'$. Using Eqs. (7) and (8) we can express the curvature tensor as

$$\kappa_{ij} = \partial_{ij}^2 d = \frac{\sqrt{2}\epsilon}{1 - \phi^2}\left(\partial_{ij}^2\phi + \frac{2\phi}{1 - \phi^2}\partial_i\phi\partial_j\phi\right). \tag{9}$$

The additive curvature, $\overline{\kappa} = \Delta d$, is

$$\overline{\kappa} = \frac{\sqrt{2}\epsilon}{(1 - \phi^2)}\left(\Delta\phi + \frac{2\phi}{1 - \phi^2}|\nabla\phi|^2\right). \tag{10}$$

From the identity $f' = 1 - f^2$ and Eq. (7), we can obtain

$$\frac{\epsilon^2}{2}|\nabla\phi|^2 = \frac{1}{4}(1 - \phi^2)^2. \tag{11}$$

At this point we define the so-called *double-well potential* as

$$W(\phi) = \frac{1}{4}(1 - \phi^2)^2. \tag{12}$$

The importance of this function stems from the fact that it has two local minima at 1 and -1, that is, in the pure phases. Finally, we express $\overline{\kappa}$ in terms of the function W as

$$\overline{\kappa} = \frac{-\sqrt{2}}{(1 - \phi^2)}\left(\frac{W'(\phi)}{\epsilon} - \epsilon\Delta\phi\right). \tag{13}$$

To continue the *diffusification* of Eq. (4) we need to obtain a relation between the velocity of the interface, V_n, and ϕ. Note that at the interface, the phase field takes the value 0. If $\mathbf{x}_{LS}$ denotes the position of the interface, then we can define a new function $\overline{\phi}(\mathbf{x}_{LS}, t)$ that is always zero. Hence, its derivative will also be zero and may be expressed as

$$\frac{d\bar{\phi}}{dt} = \nabla\phi \cdot \frac{\partial x_{LS}}{\partial t} + \partial_t\phi = 0. \tag{14}$$

We can define the velocity of the interface[4] as $v_{LS} = dx_{LS}/dt$ and the previous equation becomes

$$\partial_t\phi + v_{LS} \cdot \nabla\phi = 0, \tag{15}$$

known as the *level set equation* [23]. Equation (7) can be also written as

$$\nabla\phi = \frac{1}{\sqrt{2\epsilon}} f'\left(\frac{d}{\sqrt{2\epsilon}}\right) n_{LS}. \tag{16}$$

Introducing this result into the level set equation (15) and using the identity $f' = 1 - f^2$, we obtain

$$V_n = \frac{-\sqrt{2\epsilon}}{1 - \phi^2}\partial_t\phi. \tag{17}$$

Finally, substituting $\bar{\kappa}$ and V_n from Eqs. (13) and (17) into Eq. (4), we have derived the *phase-field equation*

$$\omega\partial_t\phi + \left(\frac{W'(\phi)}{\epsilon^2} - \Delta\phi\right) = \frac{H}{\sqrt{2}\sigma\epsilon}(1 - \phi^2)\frac{\theta - \theta_m}{\theta_m}. \tag{18}$$

Similar equations may be found, for example, in [5, 14, 34, 48].

Next, we diffusify the energy balance equation. To do that, we focus on Eqs. (1) and (2). Let us consider smooth test functions v defined on the whole domain Ω, such that $v = 0$ on $\partial\Omega$. Note that the functions v are not required to be zero on Γ_{LS}, but must be continuous there. Hence, $[\![v]\!] = 0$ at Γ_{LS}. Let us show that Eqs. (1) and (2) are equivalent, for all test functions v, to the equation

$$\frac{d}{dt}\int_\Omega ev\, dx = \int_\Omega q \cdot \nabla v\, dx. \tag{19}$$

By Reynold's theorem, the left-hand side equals

$$\frac{d}{dt}\int_{\Omega_L \cup \Omega_S} ev\, dx = \int_{\Omega_L \cup \Omega_S} v\,\partial_t e\, dx - \int_{\Gamma_{LS}} [\![e]\!]v\, V_n\, da. \tag{20}$$

---

[4]Note that V_n is the normal velocity of the interface, that is, $V_n = v_{LS} \cdot n_{LS}$.

Integrating by parts the right-hand side yields

$$\int_{\Omega_L \cup \Omega_S} q \cdot \nabla v \, dx = \int_{\Omega_L \cup \Omega_S} -v \, \text{div} \, q \, dx + \int_{\Gamma_{LS}} v [\![q]\!] \cdot n_{LS} \, da. \tag{21}$$

Note that $[\![e]\!] = \lambda$. With this, we have just proven the equivalence between Eq. (19) and Eqs. (1) and (2).

Finally, we define the smooth functions $h(\phi)$, that keeps track of the liquid phase, and $k(\phi)$. These functions satisfy $h(1) = 0$, $h(-1) = 1$, $k(1) = k_S$, and $k(-1) = k_L$. Hence, the simplest choice is

$$h(\phi) = \tfrac{1}{2}(1 - \phi), \tag{22}$$

$$k(\phi) = \tfrac{1}{2}(1 + \phi)k_S + \tfrac{1}{2}(1 - \phi)k_L. \tag{23}$$

With these functions, e and q are

$$e = C_v \theta + \lambda h(\phi), \tag{24}$$

$$q = -k(\phi)\nabla\theta. \tag{25}$$

Once we obtain smooth functions for e and q across Γ_{LS}, we can rewrite Eq. (19) as

$$\partial_t e + \text{div} \, q = 0, \tag{26}$$

which upon substitution yields

$$C_v \partial_t \theta + \lambda h'(\phi)\partial_t \phi - \text{div}\left(k(\phi)\nabla\theta\right) = 0. \tag{27}$$

Equation (27) is just a classical heat equation with an additional term that accounts for latent heat produced by the interface motion. Similar equations can be found in [5, 14, 34, 48].

The phase-field model, representing an elementary solidification theory, is finally completed by the system of PDEs composed of Eqs. (18) and (27).

2.4 The Diffuse-Interface Transition Profile: Asymptotic Analysis

Once we obtain the phase-field model of solidification given by Eqs. (18) and (27), we should prove that it indeed produces a solution where the phase field takes a hyperbolic tangent profile as $\epsilon \to 0$, as previously assumed. What follows is a heuristic proof which can be formalized using *matched asymptotic expansions* [11, 24].

We will split this analysis into two parts. First, we show that the solution quickly produces the separation of phases. Second, we show that as $\epsilon \to 0$ the solution tends to the hyperbolic tangent profile postulated in Eq. (6).

Let us rescale the time by introducing $\tau = t/\epsilon^2$. Then, Eq. (18) can be written as

$$\frac{1}{\epsilon^2}\left(\partial_\tau \phi + \frac{1}{\omega}W'(\phi)\right) = \frac{1}{\omega}\Delta\phi + \frac{H}{\sqrt{2}\omega\sigma\epsilon}(1 - \phi^2)\frac{\theta - \theta_m}{\theta_m}. \tag{28}$$

Using Landau notation and focusing on a uniform phase-field which is away from the pure phases, we can also write Eq. (28) as

$$\epsilon^{-2}\left[\partial_\tau \phi + \frac{1}{\omega}W'(\phi)\right] = O(\epsilon^{-1}) \qquad \text{as } \epsilon \to 0. \tag{29}$$

As ϵ tends to zero, the left-hand side would grow faster than the right-hand side if the term in brackets was independent from ϵ. Consequently, the term in brackets has to be of order ϵ, that is, tend to zero. Hence

$$\partial_\tau \phi \approx -\frac{1}{\omega}W'(\phi), \tag{30}$$

meaning that ϕ quickly approaches the zeros of W', which correspond to $\phi = \pm 1$. This result implies that ϕ quickly separates into the two pure phases.

To prove that ϕ adopts a hyperbolic tangent profile we will denote the zero level set of ϕ as Γ_{LS}^ϵ. The interface Γ_{LS}^ϵ should converge to Γ_{LS} as $\epsilon \to 0$. We will show now that ϕ acquires a hyperbolic tangent profile near Γ_{LS}^ϵ. In order to do that, we consider a 2D problem and re-scale the spatial variables from (x_1, x_2) to (z, s), such that

$$z(x_1, x_2, t; \epsilon) = \frac{d^\epsilon(x_1, x_2, t; \epsilon)}{\epsilon}, \tag{31}$$

where d^ϵ denotes the signed distance between a point (x_1, x_2) and the interface Γ_{LS}^ϵ. Here, we have just re-scaled the distance from any point (x_1, x_2) to $\Gamma_{LS}^\epsilon(t)$. The variable $s(x_1, x_2, t; \epsilon)$ is simply the arc length of $\Gamma_{LS}^\epsilon(t)$. We define a new function, Φ, such that $\Phi(z, s, t) = \phi(x_1, x_2, t)$. Then, in a neighborhood of Γ_{LS}^ϵ we have

$$\partial_t \phi = \partial_t \Phi + \epsilon^{-1}\partial_z \Phi \, \partial_t d^\epsilon + \partial_s \Phi \, \partial_t s, \tag{32}$$

$$\Delta\phi = \epsilon^{-2}\partial_{zz}\Phi + \epsilon^{-1}\partial_z \Phi \, \Delta d^\epsilon + \partial_{ss}\Phi \, |\nabla s|^2 + \partial_s \Phi \, \Delta s. \tag{33}$$

Finally, if we substitute these values into Eq. (18) we obtain

$$\epsilon^{-2}\left(W'(\Phi) - \partial_{zz}\Phi\right) = O(\epsilon^{-1}). \tag{34}$$

Again, the term in brackets should be $O(\epsilon)$. Hence, in a neighborhood of Γ_{LS}^{ϵ} we obtain

$$W'(\Phi) - \partial_{zz}\Phi \approx 0, \tag{35}$$

which is just a second-order differential equation whose solution takes the values ± 1 in the pure phases. This is equivalent to state that $\Phi \to \pm 1$ as $z \to \pm\infty$. As we postulated, the solution is the hyperbolic tangent function

$$\Phi(z) = \tanh\frac{z}{\sqrt{2}} = \tanh\frac{d^{\epsilon}}{\epsilon\sqrt{2}}, \tag{36}$$

which is the assumption made in the derivation of the phase-field model in Sect. 2.2.

3 General Framework of Thermomechanics and Energy Dissipation for Phase-Field Models

The goal of this section is to show how phase-field theories can be described within a rigorous thermomechanical framework. The work described in this section is based on fundamental balances such as conservation of mass, linear momentum, angular momentum, or energy. For each model, we follow the same procedure to achieve constitutive relations that rely on the fact that the stated theories must be energy dissipative. The approach taken in this section follows the work of [63]. Related work in the context of phase-field models can be found in [35] and [57].

Next, we will derive the following models within a thermomechanical framework:

- The classical Allen-Cahn equation, the canonical model for non-conserved phase dynamics.
- The Cahn-Hilliard equation, the canonical model for conserved phase dynamics.
- The Navier-Stokes-Cahn-Hilliard equations, which constitute a model for two-component immiscible flows with surface tension.

Some phase-field models of solidification can also be derived from a thermomechanical point of view. Since we have already presented a model for solidification in Sect. 2, we will not derive solidification models again using a thermomechanical approach. Note that models obtained via *diffusification* are not necessarily thermomechanically consistent. For more information about thermomechanically consistent solidification models the reader is referred to [58, 67].

3.1 The Idea Behind Thermomechanically-Consistent Phase-Field Modeling

When describing phase-field models from a thermomechanical point of view, the Helmholtz free energy, from now on denoted as Ψ, plays an essential role. The importance stems from the fact that the Helmholtz free energy depends on the phase field and its gradient, such that $\Psi(\phi, \nabla\phi)$. The constitutive relations, which define the response of the material, depend on the values of the variational derivate of the energy functional with respect to ϕ and $\nabla\phi$.

For example, let us consider the *canonical* free energy defined by

$$\Psi = G(\phi) + \frac{\epsilon^2}{2}|\nabla\phi|^2, \tag{37}$$

where G is a function of ϕ. The energy functional, also known as *Ginzburg–Landau functional*, is written as

$$\mathcal{F}^\epsilon(\phi) = \int_\Omega \Psi \, dx = \int_\Omega \left(G(\phi) + \frac{\epsilon^2}{2}|\nabla\phi|^2 \right) dx. \tag{38}$$

Its variational derivative is

$$\frac{\delta \mathcal{F}^\epsilon}{\delta\phi} = G'(\phi) - \epsilon^2 \Delta\phi. \tag{39}$$

We use the variational derivative to define the *chemical potential* $\mu = \frac{\delta \mathcal{F}^\epsilon}{\delta\phi} = G'(\phi) - \epsilon^2 \Delta\phi$. In phase-field models, constitutive relations may depend on μ and/or $\nabla\mu$.

The energy functional $\mathcal{F}^\epsilon$ may be also interpreted as

$$\mathcal{F}(\Gamma_{int}) = \psi \int_{\Gamma_{int}} da, \tag{40}$$

where Γ_{int} is a surface located in the interior of Ω and ψ a constant that accounts for the surface energy. Here, the energy functional $\mathcal{F}^\epsilon$ is a diffusification of the surface energy functional $\mathcal{F}(\Gamma_{int})$. Note that $\mathcal{F}(\Gamma_{int})$ simply associates energy to an interface proportionally to its length. We can establish a relation between $\mathcal{F}^\epsilon$ and $\mathcal{F}$ by simply considering $\mathcal{F}^\epsilon(\phi^\epsilon)$, where ϕ^ϵ is a diffusification of the interface Γ_{int} that must satisfy $G(\phi^\epsilon) = \frac{\epsilon^2}{2}|\nabla\phi^\epsilon|^2$. If we take $G(\phi) = W(\phi)$ [defined in Eq. (12)], then $\phi^\epsilon(x) = \tanh\left(\frac{d_{\Gamma_{int}}(x)}{\sqrt{2}\epsilon}\right)$, where $d_{\Gamma_{int}}(x)$ is the signed distance to Γ_{int}. We can use the *co-area formula*[5] to prove that

$$\lim_{\epsilon \to 0} \frac{1}{\epsilon} \mathcal{F}^{\epsilon}(\phi^{\epsilon}) = \mathcal{F}(\Gamma_{\text{int}}). \tag{42}$$

By taking the limit in Eq. (42), the parameter ψ can be suitably identified. This means, essentially, that phase-field energies tend to surface energies as the interface thickness ϵ approaches zero.[6]

3.2 Allen-Cahn and Cahn-Hilliard

The Allen-Cahn and Cahn-Hilliard equations are the two canonical phase-field theories for non-conserved and conserved phase dynamics, respectively [22, 59]. These equations stem from the functional presented in Eq. (38), where the function $G(\phi)$ is a double-well potential. There are several choices for this function. A common option is the double-well function defined for the solidification model in Sect. 2, i.e., $G(\phi) = W(\phi) = \frac{1}{4}(1 - \phi^2)^2$. Although this is a valid option, other possibilities, such as the logarithmic double-well, are also used. In both Allen-Cahn and Cahn-Hilliard equations the free energy is time decreasing along solutions to the equations. The difference is that the Cahn-Hilliard equation displays mass-conservation. On the other hand, the Allen-Cahn equation is non-conservative.

To set the grounds for the thermomechanical derivation of the Allen-Cahn and the Cahn-Hilliard equations we start by defining the *constitutive class*[7] for Ψ:

$$\Psi = \widehat{\Psi}(\phi, \nabla \phi). \tag{43}$$

---

[5]The co-area formula is

$$\lim_{\epsilon \to 0} \frac{1}{\epsilon} \int_{\Omega} q\left(\frac{d_{\Gamma_{\text{int}}}(x)}{\epsilon}\right) dx = \alpha_q \int_{\Gamma_{\text{int}}} da, \tag{41}$$

In this case, we use a constant value for α_q, where $\alpha_q = \int_{-\infty}^{\infty} q(z) \, dz$. The co-area formula is valid for suitably decaying functions $q(z)$, see [21, Lemma 2.1].

[6]A more mathematically rigorous relation between $\mathcal{F}^{\epsilon}$ and $\mathcal{F}$ can be determined by means of the Γ-convergence theory [2, 9, 18].

[7]The term *constitutive class* means that a certain function is allowed to depend on other variables. In this case, the constitutive class of Ψ allows it to depend on ϕ and $\nabla \phi$. Fundamental laws can be used to restrict even more the constitutive class of a function and allow only certain kinds of dependence. The term *constitutive class* is commonly employed in classical mechanics; see [37, 63].

To postulate the energy-dissipative condition we denote a subset of Ω as $\mathcal{V} \subset \Omega$ and establish the following energy dissipation law

$$\frac{\mathrm{d}}{\mathrm{d}t}\left(\int_{\mathcal{V}} \widehat{\Psi}(\phi, \nabla\phi)\mathrm{d}x\right) = \mathcal{W}(\mathcal{V}) - \mathcal{D}(\mathcal{V}). \tag{44}$$

Here, we define the *dissipation* as $\mathcal{D}(\mathcal{V})$ and we will enforce it to be non-negative for all conceivable processes. The other term in the right-hand side is the *working*, denoted as $\mathcal{W}(\mathcal{V})$. Here, $\mathcal{W}$ is associated with external forces by means of energy entering or leaving the system across the boundary $\partial\mathcal{V}$. Since we choose a subset $\mathcal{V}$ fixed in time in Eq. (44), we can introduce the temporal derivative into the integral and the left-hand side can be expressed as

$$\frac{\mathrm{d}}{\mathrm{d}t}\int_{\mathcal{V}} \Psi\,\mathrm{d}x = \int_{\mathcal{V}}\left[\partial_\phi\widehat{\Psi}\partial_t\phi + \partial_{\nabla\phi}\widehat{\Psi}\cdot\partial_t(\nabla\phi)\right]\mathrm{d}x. \tag{45}$$

At this point, we define the chemical potential μ as the variational derivative of the energy functional $\mathcal{F} = \int_{\mathcal{V}}\widehat{\Psi}\,\mathrm{d}x$, such that

$$\mu = \frac{\delta\mathcal{F}}{\delta\phi} = \partial_\phi\widehat{\Psi} - \mathrm{div}\left(\partial_{\nabla\phi}\widehat{\Psi}\right). \tag{46}$$

After integrating by parts the last term on the left-hand side of Eq. (45), we can express the energy dissipation law as

$$\frac{\mathrm{d}}{\mathrm{d}t}\int_{\mathcal{V}} \Psi\,\mathrm{d}x = \int_{\mathcal{V}} \mu\,\partial_t\phi\,\mathrm{d}x + \int_{\partial\mathcal{V}} \partial_{\nabla\phi}\widehat{\Psi}\cdot\mathbf{n}_a\partial_t\phi\,\mathrm{d}a, \tag{47}$$

where $\mathbf{n}_a$ is the unit normal vector to $\partial\mathcal{V}$. With this result we have set the basis for the derivation of the Allen-Cahn and Cahn-Hilliard equations.

3.2.1 Allen-Cahn Equation

The Allen-Cahn model represents a non-conservative system. Hence, we start by postulating the following mass balance in the system

$$\frac{\partial\phi}{\partial t} = -R, \tag{48}$$

where R represents the rate of mass exchange. The system should experience free-energy dissipation. Furthermore, we can establish the following constitutive class for R:

$$R = \widehat{R}(\phi, \nabla\phi, \mu). \tag{49}$$

One of the most important features of the phase-field method is the non-standard dependence of R on the variational derivative of the free energy. We shall now substitute the value of $\partial_t \phi$ into Eq. (47) to obtain

$$\frac{d}{dt} \int_V \Psi \, dx = \int_V -\mu R \, dx + \int_{\partial V} \partial_{\nabla \phi} \widehat{\Psi} \cdot \mathbf{n}_a \, \partial_t \phi \, da. \tag{50}$$

If we split this result into the *dissipation* and *working* terms, we obtain $\mathcal{D}(V) = \int_V \mu R \, dx$ and $\mathcal{W}(V) = \int_{\partial V} \partial_{\nabla \phi} \widehat{\Psi} \cdot \mathbf{n}_a \, \partial_t \phi \, da$. With the choice $\widehat{R}(\phi, \nabla \phi, \mu) = m(\phi)\mu$, with $m(\phi) \geq 0$, the condition $\mathcal{D}(V) \geq 0$ is satisfied. Hence, for the classical choice of the free energy functional

$$\widehat{\Psi}(\phi, \nabla \phi) = W(\phi) + \frac{\epsilon^2}{2} |\nabla \phi|^2, \tag{51}$$

where $W(\phi)$ is a double well potential, we can substitute Ψ into Eq. (46) to obtain

$$\mu = W'(\phi) - \epsilon^2 \Delta \phi. \tag{52}$$

Finally, recalling the value of R, we can combine Eqs. (48) and (52) to obtain the Allen-Cahn equation

$$\frac{\partial \phi}{\partial t} = -m(\phi)\left(W'(\phi) - \epsilon^2 \Delta \phi\right). \tag{53}$$

3.2.2 Cahn-Hilliard Equation

The Cahn-Hilliard equation is the canonical phase-field theory for conserved phase-dynamics. We postule the following mass conservation[8] equation

$$\frac{\partial \phi}{\partial t} + \operatorname{div} \mathbf{h} = 0, \tag{54}$$

where the constitutive class of $\mathbf{h}$ is given by

$$\mathbf{h} = \widehat{\mathbf{h}}(\phi, \nabla \phi, \mu, \nabla \mu). \tag{55}$$

---

[8] Mass conservation in Eq. (54), i.e., $\frac{d}{dt}(\int_\Omega \phi \, dx) = 0$, is satisfied for natural boundary conditions $\mathbf{h} \cdot \mathbf{n}_a = 0$ on $\partial\Omega$, where $\mathbf{n}_a$ is the unit normal vector to $\partial\Omega$.

Using Eq. (47) and substituting the value of $\partial_t \phi$, we obtain

$$\frac{\mathrm{d}}{\mathrm{d}t} \int_V \Psi \, \mathrm{d}x = \int_V -\mu \operatorname{div} \boldsymbol{h} \, \mathrm{d}x + \int_{\partial V} \partial_{\nabla \phi} \widehat{\Psi} \cdot \mathbf{n}_a \partial_t \phi \, \mathrm{d}a. \tag{56}$$

If we integrate by parts the first term on the right-hand side, it yields

$$\frac{\mathrm{d}}{\mathrm{d}t} \int_V \Psi \, \mathrm{d}x = \int_V \boldsymbol{h} \cdot \nabla \mu \, \mathrm{d}x + \int_{\partial V} \left(-\mu \boldsymbol{h} + \partial_{\nabla \phi} \widehat{\Psi} \partial_t \phi \right) \cdot \mathbf{n}_a \, \mathrm{d}a. \tag{57}$$

Now we are able to identify $\mathcal{D}(V)$ and $\mathcal{W}(V)$.[9] Here, the dissipation is $\mathcal{D}(V) = -\int_V \boldsymbol{h} \cdot \nabla \mu \, \mathrm{d}x$. The choice

$$\widehat{\boldsymbol{h}}(\phi, \nabla \phi, \mu, \nabla \mu) = -m(\phi) \nabla \mu \quad \text{with} \quad m(\phi) \geq 0 \tag{58}$$

satisfies the energy dissipation condition. Upon substitution, we obtain the Cahn-Hilliard equation

$$\frac{\partial \phi}{\partial t} = \operatorname{div} \left(m(\phi) \nabla \left(W'(\phi) - \epsilon^2 \Delta \phi \right) \right). \tag{59}$$

3.3 Navier-Stokes-Cahn-Hilliard

We will derive a phase-field model for immiscible two-component incompressible flow with surface tension. Similar models have been derived using different approaches. This problem has been studied from a physical, mathematical, and computational point of view. Further work can be found in [8, 16, 25, 26, 36, 43, 45, 47, 49, 53]. We start by describing a problem of a fluid composed of two immiscible components with volume fractions φ_1 and φ_2 which must verify

$$\varphi_1 + \varphi_2 = 1. \tag{60}$$

The components may have different properties, such as densities, respectively ρ_1 and ρ_2, or velocities, denoted as $\boldsymbol{u}_1$ and $\boldsymbol{u}_2$. We can define properties on the mixture. For instance, the density would be defined as $\rho = \varphi_1 \rho_1 + \varphi_2 \rho_2$ and the velocity (mass-averaged) as

$$\boldsymbol{u} = \frac{1}{\rho} \left(\varphi_1 \rho_1 \boldsymbol{u}_1 + \varphi_2 \rho_2 \boldsymbol{u}_2 \right). \tag{61}$$

---

[9]The term $\mu \boldsymbol{h}$ in Eq. (57) may be interpreted as an energy flux.

From now on, for the sake of simplicity, we assume that both components have the same density, which must be constant. Moreover, without loss of generality, we take

$$\rho = \rho_1 = \rho_2 = 1. \tag{62}$$

Mass conservation for each component implies

$$\frac{d}{dt} \int_{\mathcal{P}_t} \varphi_\alpha dx = 0; \quad \alpha = 1, 2, \tag{63}$$

where $\mathcal{P}_t$ is a set of material particles in the current configuration. By Reynold's transport theorem, Eq. (63) can be written as

$$\partial_t \varphi_\alpha + \text{div}(\varphi_\alpha \boldsymbol{u}_\alpha) = 0; \quad \alpha = 1, 2. \tag{64}$$

If we sum Eq. (64) for each α, then

$$\partial_t (\varphi_1 + \varphi_2) + \text{div}(\varphi_1 \boldsymbol{u}_1 + \varphi_2 \boldsymbol{u}_2) = 0. \tag{65}$$

Now, using Eqs. (60), (61), and (62), we obtain the following simplification for the mass conservation equation:

$$\text{div } \boldsymbol{u} = 0. \tag{66}$$

We now define the phase-field variable, ϕ, the diffusion velocity for each component, $\boldsymbol{w}$, and the phase flux, $\boldsymbol{h}$. The values of these new quantities are

$$\phi = \varphi_1 - \varphi_2, \tag{67}$$
$$\boldsymbol{w}_\alpha = \boldsymbol{u}_\alpha - \boldsymbol{u}, \tag{68}$$
$$\boldsymbol{h} = \varphi_1 \boldsymbol{w}_1 - \varphi_2 \boldsymbol{w}_2. \tag{69}$$

Subtracting Eq. (64) for each α, we obtain

$$\partial_t \phi + \text{div}(\phi \boldsymbol{u}) + \text{div } \boldsymbol{h} = 0. \tag{70}$$

Assuming that a single momentum equation governs the mixture, i.e., considering a classical mixture [61, Sec. 6.2], we can state the conservation of momentum equation

$$\dot{\boldsymbol{u}} = \text{div } \boldsymbol{T} + \boldsymbol{b}. \tag{71}$$

Here, T denotes the Cauchy stress tensor of the mixture, assumed to be symmetric,[10] b represents body forces and $\dot{u}$ is the material derivative of u; see Sect. 1.2.

So far, we have postulated the mass and linear momentum principles, and a phase-field equation. To completely define the model, we must find constitutive relations for h and T. To do that, we use the Coleman–Noll procedure [15], which provides a systematic way to ensure energy dissipation. We emphasize that the Coleman-Noll approach imposes constraints on the response of the material, not on the balance laws. We assume that the total energy of the system defined in an arbitrary region of the mixture, $\mathcal{P}_t$, can be expressed as the sum of the free energy and the kinetic energy. Recalling that $\rho = 1$, the total energy can be written as

$$\mathcal{E}(\phi, u) = \int_{\mathcal{P}_t} \left(\Psi + \frac{1}{2} |u|^2 \right) dx. \tag{72}$$

The first term on the integral is assumed to belong to the constitutive class

$$\Psi = \widehat{\Psi}(D, \phi, \nabla\phi), \tag{73}$$

where D is the symmetric part of the velocity gradient of the mixture, expressed as

$$D = \frac{1}{2}(\nabla u + \nabla u^T). \tag{74}$$

Here, ∇u^T is the transpose of ∇u. The chemical potential μ, which has been defined as the variational derivative of the energy with respect to ϕ, takes the form

$$\mu = \frac{\delta \mathcal{E}}{\delta \phi} = \partial_\phi \widehat{\Psi} - \text{div}\, \partial_{\nabla\phi} \widehat{\Psi}. \tag{75}$$

We assume that the stress tensor T and the mass flux h belong to the constitutive classes

$$T = \widehat{T}(D, \phi, \nabla\phi, \mu, \nabla\mu), \tag{76}$$

$$h = \widehat{h}(D, \phi, \nabla\phi, \mu, \nabla\mu). \tag{77}$$

---

[10]Note that, on multiphase systems, balance of angular momentum does not imply the symmetry of T [7].

The energy dissipation law is

$$\frac{d}{dt} \int_{\mathcal{P}_t} \left(\Psi + \frac{1}{2}|u|^2 \right) dx = \mathcal{W}(\mathcal{P}_t) - \mathcal{D}(\mathcal{P}_t), \tag{78}$$

Where $\mathcal{W}(\mathcal{P}_t)$ denotes the *working* and $\mathcal{D}(\mathcal{P}_t)$ the *dissipation*. Of course, we require that $\mathcal{D}(\mathcal{P}_t) \geq 0$ for all conceivable processes.

Now, we proceed to manipulate the left-hand side of Eq. (78) to obtain the values of $\mathcal{D}$ and $\mathcal{W}$. To do so, we compute separately the free energy and the kinetic energy terms. Using the Reynold's transport theorem and the divergence theorem, we obtain for the first term in the integral

$$\frac{d}{dt} \int_{\mathcal{P}_t} \Psi \, dx = \int_{\mathcal{P}_t} \partial_t \Psi \, dx + \int_{\partial \mathcal{P}_t} \Psi u \cdot n_a \, da = \int_{\mathcal{P}_t} \left(\dot{\Psi} + \Psi \operatorname{div} u \right) dx. \tag{79}$$

Making use of Eqs. (66) and (73), from Eq. (79) we obtain

$$\frac{d}{dt} \int_{\mathcal{P}_t} \Psi \, dx = \int_{\mathcal{P}_t} \left(\partial_D \Psi : \dot{D} + \partial_\phi \Psi \dot{\phi} + \partial_{\nabla \phi} \Psi \cdot (\nabla \phi)^{\cdot} \right) dx. \tag{80}$$

Note that it can also be proven that

$$(\nabla \phi)^{\cdot} = \nabla \dot{\phi} - \nabla u \nabla \phi. \tag{81}$$

If we substitute Eq. (81) into Eq. (80), we can integrate by parts the term $\nabla \dot{\phi} \, \partial_{\nabla \phi} \Psi$. After that, we use the definition of μ in Eq. (75) to obtain

$$\frac{d}{dt} \int_{\mathcal{P}_t} \Psi \, dx = \int_{\mathcal{P}_t} \left(\partial_D \Psi : \dot{D} + \mu \dot{\phi} - \partial_{\nabla \phi} \Psi \cdot \nabla u \nabla \phi \right) dx + \int_{\partial \mathcal{P}_t} \dot{\phi} \partial_{\nabla \phi} \Psi \cdot n_a \, da. \tag{82}$$

From Eq. (70) we know that $\dot{\phi} = -\operatorname{div} h$. Substituting into Eq. (82) and integrating by parts, we get

$$\frac{d}{dt} \int_{\mathcal{P}_t} \Psi \, dx = \int_{\mathcal{P}_t} \left(\partial_D \Psi : \dot{D} + \nabla \mu \cdot h - \partial_{\nabla \phi} \Psi \cdot \nabla u \nabla \phi \right) dx$$
$$+ \int_{\partial \mathcal{P}_t} \left(-\mu h + \dot{\phi} \partial_{\nabla \phi} \Psi \right) \cdot n_a \, da. \tag{83}$$

Similarly, we can compute the term involving the kinetic energy. Using the mass and momentum balance and the Reynold's transport theorem, it follows

$$\frac{d}{dt} \int_{\mathcal{P}_t} \frac{1}{2}|u|^2 dx = \int_{\mathcal{P}_t} \left(u \cdot \dot{u} + \frac{1}{2}|u|^2 \operatorname{div} u \right) dx = \int_{\mathcal{P}_t} u \cdot (\operatorname{div} T + b) \, dx. \tag{84}$$

Integrating by parts the first addend in the integral, we obtain

$$\frac{d}{dt} \int_{\mathcal{P}_t} \frac{1}{2} |u|^2 dx = \int_{\mathcal{P}_t} (-T : \nabla u + b \cdot u) \, dx + \int_{\partial \mathcal{P}_t} u \cdot T \, n_a \, da. \qquad (85)$$

Finally, the sum of Eqs. (83) and (85) gives us the value of the energy time derivative. Thus we can identify the *working*, $\mathcal{W}(\mathcal{P}_t)$, and the *dissipation*, $\mathcal{D}(\mathcal{P}_t)$, terms as

$$\mathcal{W}(\mathcal{P}_t) = \int_{\mathcal{P}_t} b \cdot u \, dx + \int_{\partial \mathcal{P}_t} \left(-\mu h \cdot n_a + \dot{\phi} \partial_{\nabla \phi} \Psi \cdot n_a + u \cdot T \, n_a \right) da, \quad (86)$$

$$\mathcal{D}(\mathcal{P}_t) = \int_{\mathcal{P}_t} \left(-\partial_D \Psi : \dot{D} - \nabla \mu \cdot h + \partial_{\nabla \phi} \Psi \cdot \nabla u \nabla \phi + T : \nabla u \right) dx. \quad (87)$$

Because $\mathcal{D}(\mathcal{P}_t) \geq 0$ for all $\mathcal{P}_t$, we require the term in parentheses in $\mathcal{D}(\mathcal{P}_t)$ to be nonnegative. There are several choices that satisfy this condition, but the simplest way is to force each of the terms to be pointwisely positive or zero. First, note that the constitutive class of Ψ does not allow its dependence on $\dot{D}$. Thus, we can only make the term $-\partial_D \Psi : \dot{D}$ pointwisely positive or zero by taking $\partial_D \Psi = 0$, which also implies that $\Psi = \widehat{\Psi}(\phi, \nabla \phi)$. The condition $\mathcal{D}(\mathcal{P}_t) \geq 0$ results now in the inequality

$$- \nabla \mu \cdot h + \partial_{\nabla \phi} \Psi \cdot \nabla u \nabla \phi + T : \nabla u \geq 0. \qquad (88)$$

If we use the identity $\partial_{\nabla \phi} \Psi \cdot \nabla u \nabla \phi = \partial_{\nabla \phi} \Psi \otimes \nabla \phi : \nabla u$, then

$$- \nabla \mu \cdot h + \partial_{\nabla \phi} \Psi \otimes \nabla \phi : \nabla u + T : \nabla u \geq 0. \qquad (89)$$

Without loss of generality, we split the Cauchy stress tensor and the velocity gradient as

$$T = T^d - pI, \qquad (90)$$

$$\nabla u = D + W, \quad \text{with} \quad W = \frac{1}{2}(\nabla u - \nabla u^T), \qquad (91)$$

where I is the identity tensor and p a scalar field that represents the mechanical pressure. The mechanical pressure may be interpreted as a Lagrange multiplier of the incompressibility constraint. Using Eqs. (90) and (91), the last term in Eq. (89) becomes

$$T : \nabla u = T : (D + W) = T : D = (T^d - pI) : D = T^d : D, \qquad (92)$$

where we used basic properties of the double-dot product[11] and the fact that $-p\boldsymbol{I} : \boldsymbol{D} = -p\,\mathrm{div}\,\boldsymbol{u} = 0$. Note that $\partial_{\nabla\phi}\Psi \otimes \nabla\phi$ is not necessarily a symmetric tensor for any Ψ. However, common and physically relevant free energies always produce symmetric tensors. Hence, we assume that $\partial_{\nabla\phi}\Psi \otimes \nabla\phi$ is a symmetric tensor to obtain

$$\partial_{\nabla\phi}\Psi \otimes \nabla\phi : \nabla\boldsymbol{u} = \partial_{\nabla\phi}\Psi \otimes \nabla\phi : \boldsymbol{D}. \tag{93}$$

Finally, the inequality stated in Eq. (89) is satisfied by taking

$$\boldsymbol{h} = -m\nabla\mu, \tag{94}$$

$$\boldsymbol{T}^d + \partial_{\nabla\phi}\Psi \otimes \nabla\phi = 2\nu\boldsymbol{D}, \tag{95}$$

where m and ν are positive functions[12] that belong to the same constitutive classes as $\boldsymbol{h}$ and $\boldsymbol{T}$, respectively. The function ν represents the viscosity of the mixture and m is the mobility. We can express the Cauchy stress tensor as

$$\boldsymbol{T} = -p\boldsymbol{I} + 2\nu\boldsymbol{D} - \partial_{\nabla\phi}\Psi \otimes \nabla\phi. \tag{96}$$

Finally, we define the free energy functional (see, e.g., [46]) $\Psi = \frac{\gamma}{\epsilon}\left(W(\phi) + \frac{\epsilon^2}{2}|\nabla\phi|^2\right)$, where γ is the surface tension and $W(\phi)$ is a double-well potential. Then, we have

$$\mu = \gamma\left(\frac{W'(\phi)}{\epsilon} - \epsilon\Delta\phi\right), \tag{97}$$

$$\partial_{\nabla\phi}\Psi = \gamma\epsilon\nabla\phi. \tag{98}$$

This completes the derivation of the theory. We can rewrite the entire model substituting the values obtained into Eqs. (66), (70), and (71), which yields

$$\frac{\partial\phi}{\partial t} + \mathrm{div}(\phi\boldsymbol{u}) - \mathrm{div}\left(m\gamma\nabla\left[\frac{W'(\phi)}{\epsilon} - \epsilon\Delta\phi\right]\right) = 0, \tag{99}$$

$$\dot{\boldsymbol{u}} + \nabla p = \mathrm{div}\left(2\nu\boldsymbol{D} - \gamma\epsilon\nabla\phi \otimes \nabla\phi\right) + \boldsymbol{b}, \tag{100}$$

$$\mathrm{div}\,\boldsymbol{u} = 0. \tag{101}$$

---

[11] If A is a symmetric tensor, then $A : B = A : (B + B^T)/2$. Consequently, it can easily be proven that if A is symmetric and B is skew-symmetric, then $A : B = 0$.

[12] Usually, m and ν are assumed to be constant.

If we substitute the expression of Ψ in Eq. (72), the energy of the system $\mathcal{P}_t$ may be written as

$$\mathcal{E}(\phi, \boldsymbol{u}) = \int_{\mathcal{P}_t} \left(\frac{1}{2}|\boldsymbol{u}|^2 + \gamma \left[\frac{W(\phi)}{\epsilon} + \frac{\epsilon}{2}|\nabla\phi|^2 \right] \right) dx. \tag{102}$$

The free energy in (102) can be interpreted in the limit $\epsilon \to 0$ as the sum of kinetic energy and surface energy. This can be understood by examining Eqs. (38), (40), and (42) and comparing them with the term in brackets in Eq. (102).

4   The Diffuse Domain Approach

A great number of problems in different areas require solving PDEs on moving domains. These problems entail certain complications such as the need of moving meshes or the explicit tracking of the interface. There are different approaches to solve this kind of problems, for instance, interface capturing methods or boundary integral methods [40, 60]. In this section, we show how the diffuse domain approach can be used to tackle the difficulties that stem from moving domains. The diffuse domain method allows us to solve the PDEs using a fixed underlying mesh. This method is strictly related to phase-field modeling, since it requires the use of a variable as a marker of the location of the moving domain. The phase field acquires the value 1 inside the moving domain, and 0 outside. Consequently, the moving computational domain, $\Omega_\rho(t)$, has to be extended to a larger one, named Ω, which contains the former, i.e., $\Omega_\rho(t) \subset \Omega$ for all t. The moving boundary is now represented by a quick but smooth transition between the two values of the phase field. In addition, the diffuse domain approach is often used to solve PDEs on fixed domains. When the geometry is complex, this approach can be useful to avoid meshing domains with complicated topologies.

4.1   Derivation of a Diffuse Domain Model for a Standard Convection-Diffusion Problem

We proceed to derive the diffuse domain approach for a convection-diffusion-reaction problem posed on a moving domain. Let us consider the concentration of a generic compound, ρ, which is subject to convection, reaction and diffusion within a domain Ω_ρ that changes with time, i.e., $\Omega_\rho = \Omega_\rho(t)$. The governing equation may be written as

$$\partial_t \rho + \text{div}(\rho \boldsymbol{u}_\rho) = \text{div}(D\nabla\rho) + R \quad \text{in} \quad \Omega_\rho(t), \tag{103}$$

where $\boldsymbol{u}_\rho$ is the velocity that transports ρ, D is the diffusion coefficient, assumed to be constant, and R represents a generic reactive term. We impose a flux across $\Gamma_\rho = \Gamma_\rho(t)$, the boundary of Ω_ρ. Then, the boundary condition is

$$D\nabla\rho \cdot \boldsymbol{n}_{\Gamma_\rho} + \rho(\boldsymbol{u}_{\Gamma_\rho} - \boldsymbol{u}_\rho) \cdot \boldsymbol{n}_{\Gamma_\rho} = -j \quad \text{in} \quad \Gamma_\rho(t), \tag{104}$$

where $\boldsymbol{u}_{\Gamma_\rho}$ is the velocity of the interface (in general, different from $\boldsymbol{u}_\rho$), $\boldsymbol{n}_{\Gamma_\rho}$ is the unit normal vector to Γ_ρ, and j is a prescribed value of the flux. To start the derivation of the model, we multiply Eq. (103) by a smooth function w and integrate in space and time. Hence, Eq. (103) becomes

$$\int_0^T \int_{\Omega_\rho} w\, \partial_t \rho \, dx\, dt + \int_0^T \int_{\Omega_\rho} w\, \text{div}(\rho \boldsymbol{u}_\rho)\, dx\, dt =$$
$$\int_0^T \int_{\Omega_\rho} w\, \text{div}(D\nabla\rho)\, dx\, dt + \int_0^T \int_{\Omega_\rho} wR\, dx\, dt. \tag{105}$$

We choose weight functions w that are time independent. Thus, we can introduce w in the time derivative and rewrite the first term in Eq. (105) as

$$\int_0^T \int_{\Omega_\rho} \partial_t(w\rho)\, dx\, dt = -\int_0^T \int_{\Gamma_\rho} w\rho\, \boldsymbol{u}_{\Gamma_\rho} \cdot \boldsymbol{n}_{\Gamma_\rho}\, da\, dt + \int_{\Omega_\rho(T)} w\rho\, dx - \int_{\Omega_\rho(0)} w\rho\, dx. \tag{106}$$

Equation (106) can be derived using the Reynolds transport theorem to transform the integral on Ω_ρ on an integral on a time independent domain. Using standard identities from continuum mechanics, Eq. (106) follows. We next integrate by parts the second term on the left-hand side and the first term on the right-hand side in Eq. (105), such that

$$\int_0^T \int_{\Omega_\rho} w\, \text{div}(\rho \boldsymbol{u}_\rho)\, dx dt = -\int_0^T \int_{\Omega_\rho} \rho\, \nabla w \cdot \boldsymbol{u}_\rho\, dx dt + \int_0^T \int_{\Gamma_\rho} w\, \rho\, \boldsymbol{u}_\rho \cdot \boldsymbol{n}_{\Gamma_\rho}\, da\, dt. \tag{107}$$

$$\int_0^T \int_{\Omega_\rho} w\, \text{div}(D\nabla\rho)\, dx\, dt = -\int_0^T \int_{\Omega_\rho} D\nabla w \cdot \nabla\rho\, dx\, dt + \int_0^T \int_{\Gamma_\rho} w\, D\nabla\rho \cdot \boldsymbol{n}_{\Gamma_\rho}\, da\, dt. \tag{108}$$

Now, we make use of a phase-field property to replace the integrals in $\Omega_\rho(t)$ and its boundary $\Gamma_\rho(t)$ by integrals in a larger and fixed domain Ω, such that $\Omega_\rho(t) \subset \Omega$. This property allows us to change the integration domain by simply multiplying the integrand by a smooth function that localizes the domain Ω_ρ or its interface Γ_ρ. We are able to localize $\Omega_\rho(t)$ with the function ϕ, and its boundary $\Gamma_\rho(t)$ with the function $\delta_\Gamma(\phi)$, which takes the value 1 in Γ_ρ and 0 elsewhere.[13] Using this, we can write the integrals as

$$\int_{\Omega_\rho} F \, dx = \int_\Omega \phi \, F \, dx \quad \text{and} \quad \int_{\Gamma_\rho} F \, da = \int_\Omega \delta_\Gamma F \, dx, \tag{109}$$

where F is a generic function. Using Eqs. (104) and (106)–(109), we can rewrite Eq. (105) as

$$\int_\Omega \phi(x, T) \, w \, \rho \, dx - \int_\Omega \phi(x, 0) \, w \, \rho \, dx - \int_0^T \int_\Omega \phi \rho \nabla w \cdot \boldsymbol{u}_\rho \, dx \, dt + \int_0^T \int_\Omega \delta_\Gamma \, w \, j \, dx \, dt =$$

$$- \int_0^T \int_\Omega \phi \, D \nabla w \cdot \nabla \rho \, dx \, dt + \int_0^T \int_\Omega \phi \, w \, R \, dx \, dt. \tag{110}$$

Integrating by parts the terms containing ∇w and manipulating the first two terms in Eq. (110), we obtain

$$\int_0^T \int_\Omega w \, \partial_t(\phi\rho) \, dx \, dt + \int_0^T \int_\Omega w \, \text{div}(\phi\rho\boldsymbol{u}_\rho) \, dx \, dt - \int_0^T \int_\Gamma w \, \phi\rho \, \boldsymbol{u}_\rho \cdot \boldsymbol{n}_\Gamma \, da \, dt =$$

$$\int_0^T \int_\Omega w \, \text{div}(\phi D \nabla \rho) \, dx \, dt - \int_0^T \int_\Gamma w \, \phi D \nabla \rho \cdot \boldsymbol{n}_\Gamma \, da \, dt$$

$$+ \int_0^T \int_\Omega w \, \phi R \, dx \, dt - \int_0^T \int_\Omega w \, \delta_\Gamma j \, da \, dt, \tag{111}$$

where $\boldsymbol{n}_\Gamma$ is the unit normal vector to $\Gamma = \partial\Omega$. For simplicity, we consider periodic boundary conditions in Ω and, thus, the integrals on Γ vanish. The resulting equation may be written as

$$\partial_t(\phi\rho) + \text{div}(\phi\rho\boldsymbol{u}_\rho) = \text{div}(\phi D \nabla \rho) + \phi R - \delta_\Gamma j \quad \text{in} \quad \Omega, \tag{112}$$

which is a diffusion-convection-reaction equation posed on the fixed domain Ω. This equation has to be coupled with a phase-field equation that controls the dynamics of the phase field, i.e., that accounts for the motion of the domain Ω_ρ. Here, we use the phase-field equation

$$\partial_t\phi + \boldsymbol{u}_{\Gamma_\rho} \cdot \nabla\phi = \Gamma_\phi \left(\epsilon \Delta\phi - \frac{W'(\phi)}{\epsilon} + \kappa\epsilon|\nabla\phi| \right), \tag{113}$$

---

[13]The smooth function δ_Γ works as a marker of the position of the boundary of the moving domain. A suitable expression for this function could be, e.g., $\delta_\Gamma = \epsilon^2|\nabla\phi|^2$, though there are other valid expressions for δ_Γ in the literature.

where ϵ is a parameter that controls the thickness of the interface, Γ_ϕ is a parameter that sets the strength of the right-hand side, and $\kappa = -\text{div}\,(\nabla\phi/|\nabla\phi|)$ is the curvature of the interface. The function $W(\phi)$ is a double well potential with local minima at 0 and 1. The right-hand side of the equation imposes a hyperbolic tangent profile on the interface, while the term containing u_{Γ_ρ} accounts for the movement of the domain $\Omega_\rho(t)$. A similar equation can be found in [4].

The system composed of Eqs. (112) and (113) is the so-called diffuse domain problem, which is equivalent to the problem defined in Eqs. (103) and (104) as $\epsilon \to 0$. The diffuse domain problem is posed on a bigger and fixed computational domain and accounts for the motion of the domain Ω_ρ and the dynamics of the compound ρ within Ω_ρ.

4.2 Diffuse Domain Method for Problems Posed on Evolving Surfaces

We can also apply the diffuse domain approach to problems that involve the solution of PDEs on moving surfaces. Let us consider a generic substance located in a moving surface $\Gamma_\rho(t)$. The substance, with concentration $\rho_\Gamma(x, t)$, undergoes diffusion, reaction and convection within the surface Γ_ρ. The PDE that controls the convection-diffusion-reaction process may be written as

$$\partial_t \rho_\Gamma + \text{div}_\Gamma(\rho_\Gamma u_\Gamma) = \text{div}_\Gamma(D_\Gamma \nabla_\Gamma \rho_\Gamma) + R_\Gamma \quad \text{in} \quad \Gamma_\rho, \tag{114}$$

where $\text{div}_\Gamma(\cdot)$ and ∇_Γ denote, respectively, the operators $\text{div}(\cdot)$ and ∇ applied on the surface Γ_ρ. In Eq. (114), D_Γ is the diffusion coefficient, u_Γ is the velocity of the compound,[14] and R_Γ is a generic reactive term.

The procedure to obtain the equivalent phase-field equation is similar to that described in Sect. 4.1. However, the derivation is more involved because we are transforming a 2D problem (posed on a surface) into a 3D problem. Here, we locate the surface Γ_ρ with the function $\delta_\Gamma(\phi)$, introduced in Sect. 4.1. Thus, we use the function δ_Γ to change the integration domain from Γ_ρ to Ω. Following a similar procedure to that explained before, we obtain

$$\partial_t(\delta_\Gamma \rho_\Gamma) + \text{div}(\delta_\Gamma \rho_\Gamma u_\Gamma) = \text{div}(\delta_\Gamma D_\Gamma \nabla \rho_\Gamma) + \delta_\Gamma R_\Gamma \quad \text{in} \quad \Omega, \tag{115}$$

where Ω is a fixed domain such that $\Gamma_\rho(t) \subset \Omega$ for all t. Equation (115) is the diffuse domain equation equivalent to Eq. (114). To complete the diffuse domain model, we need to couple Eq. (115) with another equation (for example, Eq. (113)) that accounts for the phase field dynamics.

---

[14]Note that the velocity of ρ_Γ within the surface Γ_ρ (i.e., u_Γ) does not necessarily coincide with the velocity of the surface (u_{Γ_ρ}).

5 Numerical Examples

In this section we illustrate how the phase-field models proposed in the previous sections can be numerically solved. In Sect. 5.1 we solve the solidification model, in Sect. 5.2 we focus on the classical Cahn-Hilliard equation, and in Sect. 5.3 we test the diffuse domain approach with a diffusion problem posed on a moving domain.

5.1 Solidification Model

We start with a similar mathematical model to that proposed in Sect. 2.3, and the we will obtain the weak formulation of the problem. The system of PDEs may be written as

$$\omega\epsilon\,\partial_t\phi + \left(\frac{W'(\phi)}{\epsilon} - \epsilon\Delta\phi\right) = \frac{H}{\sqrt{2\sigma}}\left(1-\phi^2\right)^2\frac{\theta - \theta_m}{\theta_m}, \tag{116}$$

$$C_v\partial_t\theta + \lambda h'(\phi)\partial_t\phi - \operatorname{div}\left(k(\phi)\nabla\theta\right) = 0. \tag{117}$$

While Eqs. (117) and (27) are identical, Eq. (116) displays some differences with respect to Eq. (18). On the right-hand side of Eq. (116), we use the function $\left(1-\phi^2\right)^2$ instead of $\left(1-\phi^2\right)$. While from a qualitative point of view $(1-\phi^2)$ and $(1-\phi^2)^2$ accomplish the same goal [localizing the right-hand side of Eq. (116) to the interface], the squared term fosters dendritic growth and leads to solutions that mimic real solidification patterns. Essentially, Eq. (116) can be regarded as an Allen-Cahn equation, where the right-hand side is a reactive term that drives the phase transformation depending on the value of the temperature θ. The mass transfer between the phases is located in the neighborhood of the interface through the function $\left(1-\phi^2\right)^2$, which vanishes in the phases $\phi = \pm1$. A similar phase-field model of solidification can be found in [1].

We derive a weak formulation by considering smooth functions η, w, which belong to the Sobolev space of square-integrable functions with square-integrable first derivatives, i.e., $\mathcal{V} = \{\eta, w \mid \eta, w \in \mathcal{H}^1(\Omega)\}$. Multiplying Eq. (116) by η, Eq. (117) by w and integrating in the domain Ω yields

$$0 = \int_\Omega \eta\omega\epsilon\,\partial_t\phi\,dx + \int_\Omega \eta\frac{W'(\phi)}{\epsilon}\,dx - \int_\Omega \eta\,\epsilon\Delta\phi\,dx - \int_\Omega \eta\,\frac{H}{\sqrt{2\sigma}}\left(1-\phi^2\right)^2\frac{\theta - \theta_m}{\theta_m}\,dx, \tag{118}$$

$$0 = \int_\Omega wC_v\partial_t\theta\,dx + \int_\Omega w\lambda h'(\phi)\partial_t\phi\,dx - \int_\Omega w\operatorname{div}\left(k(\phi)\nabla\theta\right)dx, \tag{119}$$

where we considered periodic boundary conditions to simplify the formulation. After integrating by parts, we can express the variational problem as: Find $\phi, \theta \in \mathcal{V}$ such that for all $\eta, w \in \mathcal{V}$

$$0 = \int_\Omega \eta \omega \epsilon \partial_t \phi \, dx + \int_\Omega \eta \frac{W'(\phi)}{\epsilon} \, dx + \int_\Omega \epsilon \nabla \eta \cdot \nabla \phi \, dx - \int_\Omega \eta \frac{H}{\sqrt{2}\sigma} \left(1 - \phi^2\right)^2 \frac{\theta - \theta_m}{\theta_m} \, dx,$$

(120)

$$0 = \int_\Omega w \, C_v \partial_t \theta \, dx + \int_\Omega w \lambda h'(\phi) \partial_t \phi \, dx + \int_\Omega k(\phi) \nabla w \cdot \nabla \theta \, dx.$$

(121)

Now, by using isogeometric analysis we can construct a discrete space $\mathcal{V}^h \subset \mathcal{V}$ comprised of B-Splines or NURBS. Using the Galerkin method, the problem may be stated as: Find $\phi^h, \theta^h \in \mathcal{V}^h$ such that for all $\eta^h, w^h \in \mathcal{V}^h$

$$0 = \int_\Omega \eta^h \omega \epsilon \partial_t \phi^h \, dx + \int_\Omega \eta^h \frac{W'(\phi^h)}{\epsilon} \, dx + \int_\Omega \epsilon \nabla \eta^h \cdot \nabla \phi^h \, dx$$
$$+ \int_\Omega \eta^h \frac{H}{\sqrt{2}\sigma} \left(1 - \left(\phi^h\right)^2\right)^2 \frac{\theta^h - \theta_m}{\theta_m} \, dx,$$

(122)

$$0 = \int_\Omega w^h \, C_v \partial_t \theta^h \, dx + \int_\Omega w^h \lambda h'(\phi^h) \partial_t \phi^h \, dx + \int_\Omega k(\phi^h) \nabla w^h \cdot \nabla \theta^h \, dx,$$

(123)

which is the semidiscrete formulation of the solidification model. To perform the time discretization we use the generalized-α method [44]. Details about the implementation of the generalized-α method for a phase-field equation can be found in [30].

What is left is defining the parameters and functions in the model. Our goal is to reproduce the growth of an undercooled crystal of pure nickel. The properties of the material are $C_v = 5.42 \, \mathrm{J\,K^{-1}\,cm^{-3}}$, $\lambda = 2450 \, \mathrm{J\,cm^{-3}}$, $k_S = k_L = 0.8401 \, \mathrm{J\,K^{-1}\,cm^{-1}\,s^{-1}}$, $\sigma = 3.7 \cdot 10^{-5} \mathrm{J\,cm^{-2}}$, $\omega = 130 \mathrm{s\,cm^{-2}}$, $H = -4.2336 \mathrm{J\,cm^{-3}}$, $\theta_m = 1728 \, \mathrm{K}$, and $\epsilon = 2 \cdot 10^{-4} \mathrm{cm}$. We assume an initial temperature of θ_m for the crystal and $\theta_m - 217 \, \mathrm{K}$ for the liquid phase. The crystal's initial shape is irregular to trigger dendritic growth at early stages of the simulation. The functions required to completely define the model are those described in Sect. 2.3, namely,

$$W(\phi) = \tfrac{1}{4} \left(1 - \phi^2\right)^2,$$

(124)

$$h(\phi) = \tfrac{1}{2}(1 - \phi),$$

(125)

$$k(\phi) = \tfrac{1}{2}(1 + \phi)k_S + \tfrac{1}{2}(1 - \phi)k_L.$$

(126)

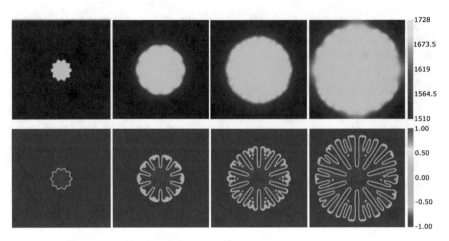

Fig. 1 Time evolution of the temperature (top row) and the phase-field (bottom row). We use a square computational domain of side $L = 0.053$ cm meshed with 400^2 quadratic elements, which leads to an overkilled solution. As initial condition we take a small irregular crystal of pure nickel with an average radius of 0.0053 cm

Figure 1 shows the time evolution of the temperature and the phase field. We observe early dendritic growth followed by splitting of dendrites. Due to the phase transformation, the solution along lines perpendicular to the interface deviates slightly from the hyperbolic-tangent profile on the tips of the growing dendrites. The images show that interface regions where the crystal is growing have a larger interfacial length. This feature is even more remarkable in simulations performed with the original model derived in Sect. 2.3, where the function that localizes the interface is simply quadratic (not shown). However, the hyperbolic-tangent profile is recovered as the crystal stops growing.

Even more realistic and complex dendritic patterns can be obtained if we introduce anisotropy in the model. This can be done by assuming anisotropic surface tension, which means that σ is now a function of the unit normal to the solid-liquid interface. We use the function $\sigma/\bar{\sigma} = 1 + \delta \cos[q(\alpha - \alpha_0)]$, where $\bar{\sigma}$ is the average surface tension, δ is the strength of the anisotropy and q and α_0 are the mode number and preferred angle respectively. The angle of the normal to the liquid-solid interface can be easily defined from the phase-field. The simulations shown in Fig. 2 were performed with the same parameters as before, except for the enthalpy, that now has a value of $H = -42.336 \mathrm{J\,g}^{-1}$. The parameters that define anisotropy are $\delta = 0.05$, $\alpha_0 = \pi/4$, $\bar{\sigma} = 3.7 \cdot 10^{-5} \mathrm{J\,cm}^{-2}$, and $q = 4$. Also, the initial condition does not need to have large irregularities because an almost perfect circular shape is able to trigger dendritic growth due to anisotropy. Note that, although almost imperceptible, there are small irregularities on the initial shape of the crystal.

Fig. 2 Time evolution of the temperature (top row) and the phase-field (bottom row) in an anisotropic case. We use a square computational domain of side $L = 0.1$ cm meshed with 512^2 quadratic elements. As initial condition we take a small irregular crystal of pure nickel with an average radius of 0.00264 cm

5.2 Cahn-Hilliard Equation

We will derive the weak formulation of the Cahn-Hilliard problem introduced in Sect. 3.2. The Cahn-Hilliard equation is

$$\frac{\partial \phi}{\partial t} = \text{div}\left(m(\phi)\nabla\left(W'(\phi) - \epsilon^2 \Delta\phi\right)\right) \quad \text{in} \quad \Omega. \tag{127}$$

For simplicity we consider periodic boundary conditions. We follow a similar procedure as in Sect. 5.1, but now we require w to belong to the Sobolev space of square-integrable functions with square-integrable second derivatives, i.e., $\mathcal{V} = \{w \mid w \in \mathcal{H}^2(\Omega)\}$, to obtain the following variational problem: Find $\phi \in \mathcal{V}$ such that for all $w \in \mathcal{V}$

$$0 = \int_\Omega w \frac{\partial \phi}{\partial t}\, dx + \int_\Omega \nabla w \cdot \nabla\phi\, W''(\phi)\, m(\phi)\, dx$$

$$+ \int_\Omega \nabla w \cdot \nabla\phi\, m'(\phi)\epsilon^2 \Delta\phi\, dx + \int_\Omega \Delta w\, \epsilon^2 m(\phi)\Delta\phi\, dx. \tag{128}$$

Again, we resort to isogeometric analysis to construct a discrete space $\mathcal{V}^h \subset \mathcal{V}$, where we use B-Splines or NURBS of order $p \geq 2$ and maximum smoothness. The semidiscrete form of the Cahn-Hilliard equation is stated as: Find $\phi^h \in \mathcal{V}^h$ such that for all $w^h \in \mathcal{V}^h$

$$
0 = \int_\Omega w^h \frac{\partial \phi^h}{\partial t} \, dx + \int_\Omega \nabla w^h \cdot \nabla \phi^h W''(\phi^h) m(\phi^h) \, dx
$$

$$
+ \int_\Omega \nabla w^h \cdot \nabla \phi^h m'(\phi^h) \epsilon^2 \Delta \phi^h \, dx + \int_\Omega \Delta w^h m(\phi^h) \epsilon^2 \Delta \phi^h \, dx. \tag{129}
$$

The time discretization was performed using the generalized-α method. In addition, since the simulations display a wide range of time scales [27, 30], we have implemented an adaptive time-stepping scheme to reduce the computational time.

We now define the specific functions of the model which control the dynamics of the simulations. The mobility is defined as

$$
m(\phi) = \frac{1}{4} \left(1 - \phi^2 \right). \tag{130}
$$

Although sometimes the mobility is assumed constant, this form of the mobility, also known as degenerate mobility, is more fundamental from the point of view of mixture chemistry. The most remarkable feature of this function is that $m(\phi)$ vanishes in the pure phases, which means that the motion is restricted to the interface. Next, we define the double-well potential. In this case, we use the logarithmic double well potential, defined as

$$
W(\phi) = \frac{1}{2} \left[(1 + \phi) \log \left(\frac{1 + \phi}{2} \right) + (1 - \phi) \log \left(\frac{1 - \phi}{2} \right) + \theta (1 - \phi^2) \right]. \tag{131}
$$

The parameter θ in the last term of Eq. (131) represents the quench temperature and takes the value $\theta = 3/2$ in our simulations.

Using dimensional analysis, it may be proven that the solution of the Cahn-Hilliard equation only depends on the dimensionless number [30]

$$
\alpha = \frac{L_0^2}{3\epsilon^2}, \tag{132}
$$

where L_0 is a length scale of the problem. We assume $L_0 = 1$ to be the size of the computational domain.

We perform two simulations with slightly different initial conditions, where we can observe the processes involved in the Cahn-Hilliard equation. The first simulation starts with values of the phase field randomly chosen around a fixed initial value. Thus, the initial condition takes the form $\phi(x, 0) = \bar{\phi} + \eta(x)$, where $\bar{\phi}$ is a constant and $\eta(x)$ is a small perturbation. If $\bar{\phi}$ is in the spinodal region, that is, $W''(\bar{\phi}) < 0$ (see [33]), the initial condition is unstable and the perturbation grows in time, eventually leading to phase separation. Figure 3 shows snapshots of the simulation at four different times. The phase separation manifests itself in the form of patches, or bubbles, of one of the phases distributed inside the other. The bubble

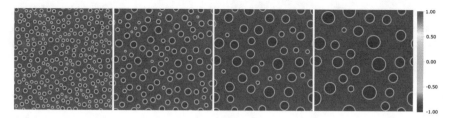

Fig. 3 2D Cahn-Hilliard simulation performed in the domain $\Omega = [0, 1]^2$. The phase field quickly separates into the two phases (blue and red). As time evolves (left to right), the system *coarsens*. The mesh is uniform and composed of 256^2 C^1-quadratic elements. Here, $\alpha = 10^5$

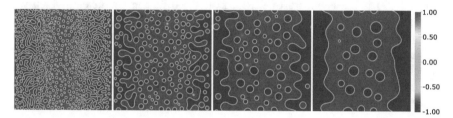

Fig. 4 2D Cahn-Hilliard simulation performed in the domain $\Omega = [0, 1]^2$. The mixture undergoes two different processes for phase separation, (1) bubble merging in the vertical borders and central vertical axis, and (2) spinodal decomposition between those regions. Again, we use an uniform mesh composed of 256^2 C^1-quadratic elements and $\alpha = 10^5$

phase and the dominating phase take the values $\phi \approx \pm 1$ with a quick transition between them. After the initial phase separation, the process of *coarsening* starts. Smaller areas tend to disappear and merge with larger ones, producing fewer but larger bubbles to reduce the energy of the system.

Figure 4 shows the time evolution of a simulation where $\phi(x, 0) = -4x^2 + 4x - 0.5 + \eta(x)$. This implies that the initial values for the phase-field are $\phi \approx 0.5$ in the center of the domain, and $\phi \approx -0.5$ in the borders if we move in the x-direction. We can observe, in addition to the coarsening process, the so-called spinodal decomposition [33], which tends to produce a stripped pattern. Time evolution of the phase field shows bubble merging near the left and right vertical boundaries and the region around the vertical line at $x = 0.5$. However, in the borders, the phase $\phi = -1$ (blue) dominates while in the center is the phase $\phi = 1$ (red) the one that dominates. Hence, accounting for mass conservation, the solution tends to accumulate the blue phase in the vertical borders and the red phase in between.

5.3 Diffuse Domain Model

Here, we show how the diffuse domain approach can be used to obtain accurate solutions of convection-diffusion problems. In particular, we use the sharp-interface and the diffuse-interface approaches to solve the same problem and then, compare the solutions.

The sharp-interface approach is referred to as the *reference* problem. Let us consider a compound of concentration $\rho^{\text{ref}}(x, t)$, subject to diffusion and living in a circular moving domain $\Omega_\rho(t)$ of radius R_0. We assume that $\Omega_\rho(t)$ moves with constant velocity u_{Γ_ρ}. The governing equations can be written as

$$\partial_t \rho^{\text{ref}} + \text{div}(\rho^{\text{ref}} u_{\Gamma_\rho}) - \text{div}\left(D\nabla\rho^{\text{ref}}\right) = 0 \text{ in } \Omega_\rho(t), \tag{133}$$

$$D\nabla\rho^{\text{ref}} \cdot n_{\Gamma_\rho} = 0 \text{ on } \Gamma_\rho(t), \tag{134}$$

$$\rho^{\text{ref}}(x, 0) = \rho_0(x) \text{ in } \overline{\Omega}_\rho(0), \tag{135}$$

where $\Gamma_\rho(t)$ is the boundary of $\Omega_\rho(t)$, D is the diffusion coefficient, which we assume constant, n_{Γ_ρ} is the unit normal pointing outwards Γ_ρ, and $\rho_0(x)$ is the initial concentration of ρ^{ref}. Note that Eq. (134) is equivalent to Eq. (104), where we took $u_\rho = u_{\Gamma_\rho}$ and $j = 0$. Since we assumed that $\Omega_\rho(t)$ moves with constant velocity u_{Γ_ρ}, we can solve the reference problem in a fixed domain $\Omega_0 = \Omega_\rho(0)$ and translate the solution. The problem can be reformulated as a function of the concentration $\hat{\rho}(x, t)$ as

$$\partial_t \hat{\rho} - \text{div}\left(D\nabla\hat{\rho}\right) = 0 \text{ in } \Omega_0, \tag{136}$$

$$\nabla\hat{\rho} \cdot n_{\Gamma_0} = 0 \text{ on } \Gamma_0, \tag{137}$$

$$\hat{\rho}(x, 0) = \rho_0(x) \text{ in } \overline{\Omega}_0, \tag{138}$$

such that $\rho^{\text{ref}}(x, t) = \hat{\rho}(x - u_{\Gamma_\rho}t, t)$, where Γ_0 is the boundary of Ω_0 and n_{Γ_0} is its unit normal vector.

If we follow the same procedure as in Sect. 4, from Eqs. (133)–(135) we obtain the equivalent diffuse-interface problem:

$$\partial_t \phi + u_{\Gamma_\rho} \cdot \nabla\phi = \Gamma_\phi \left(\epsilon\Delta\phi - \frac{W'(\phi)}{\epsilon} + \kappa\epsilon|\nabla\phi|\right) \text{ in } \Omega, \tag{139}$$

$$\partial_t (\phi\rho_\phi) + \text{div}(\phi\rho_\phi u_{\Gamma_\rho}) = \text{div}(D\phi\nabla\rho_\phi) \text{ in } \Omega, \tag{140}$$

$$\phi(x, 0) = \phi_0(x) \text{ in } \overline{\Omega}, \tag{141}$$

$$\rho_\phi(x, 0) = \rho_{\phi 0}(x) \text{ in } \overline{\Omega}, \tag{142}$$

where $\rho_\phi(x, t)$ is the diffuse-domain approximation to $\rho^{\text{ref}}(x, t)$ and ϕ is the phase field, which determines the position of $\Omega_\rho(t)$. We use the double-well potential $W(\phi) = 18\phi^2(1 - \phi)^2$. The initial position of Ω_ρ is represented by $\phi_0(x)$, defined as $\phi_0(x) = 0.5 + 0.5\tanh[\frac{2\sqrt{2}}{\epsilon}(d(x) - R_0)]$, where R_0 is the radius of Ω_ρ and $d(x)$ the distance from point x to the center of the domain. Finally, $\rho_{\phi 0}(x)$ is the initial concentration of ρ^{ref}, defined in the entire domain Ω such that $\rho_{\phi 0}(x) = \rho_0(x)$ in the region occupied by Ω_ρ and $\rho_{\phi 0}(x) = 0$ elsewhere.

Following the same procedure as in previous sections, now with w, η belonging to the Sobolev space of square-integrable functions with square-integrable second derivatives, i.e., $\mathcal{V} = \{w, \eta \mid w, \eta \in \mathcal{H}^2(\Omega)\}$, we obtain the variational problem corresponding to the diffuse-interface problem[15] assuming periodic boundary conditions: Find ϕ, $\rho_\phi \in \mathcal{V}$ such that for all η, $w \in \mathcal{V}$

$$0 = \int_\Omega \eta \frac{\partial \phi}{\partial t} \, dx + \int_\Omega \eta \, u_{\Gamma_\rho} \cdot \nabla\phi \, dx + \int_\Omega \Gamma_\phi \epsilon \nabla\eta \cdot \nabla\phi \, dx$$

$$+ \int_\Omega \eta \, \Gamma_\phi \frac{W'(\phi)}{\epsilon} \, dx - \int_\Omega \Gamma_\phi \epsilon \nabla\eta \cdot \nabla\phi \, dx - \int_\Omega \eta \frac{\Gamma_\phi \epsilon}{|\nabla\phi|} \nabla(|\nabla\phi|) \cdot \nabla\phi \, dx,$$

$$(143)$$

$$0 = \int_\Omega w \frac{\partial(\phi\rho_\phi)}{\partial t} dx - \int_\Omega \phi\rho_\phi \nabla w \cdot u_{\Gamma_\rho} \, dx + \int_\Omega D\phi \nabla w \cdot \nabla\rho_\phi \, dx, \qquad (144)$$

Equation (143) does not represent the simplest weak form of Eq. (139), but one that is convenient for numerical calculations; more details can be found in [56]. Using isogeometric analysis, we construct a discrete space $\mathcal{V}^h \subset \mathcal{V}$ making use of maximum-continuity B-Splines or NURBS of order $p \geq 2$. The semidiscrete form of the problem is written as: Find ϕ^h, $\rho_\phi^h \in \mathcal{V}^h$ such that for all η^h, $w^h \in \mathcal{V}^h$

$$0 = \int_\Omega \eta^h \frac{\partial \phi^h}{\partial t} dx + \int_\Omega \eta^h \, u_{\Gamma_\rho} \cdot \nabla\phi^h \, dx + \int_\Omega \Gamma_\phi \epsilon \nabla\eta^h \cdot \nabla\phi^h \, dx$$

$$+ \int_\Omega \eta^h \, \Gamma_\phi \frac{W'(\phi^h)}{\epsilon} \, dx - \int_\Omega \Gamma_\phi \epsilon \nabla\eta^h \cdot \nabla\phi^h \, dx - \int_\Omega \eta^h \frac{\Gamma_\phi \epsilon}{|\nabla\phi^h|} \nabla(|\nabla\phi^h|) \cdot \nabla\phi^h \, dx,$$

$$(145)$$

$$0 = \int_\Omega w^h \frac{\partial(\phi^h \rho_\phi^h)}{\partial t} dx - \int_\Omega \phi^h \rho_\phi^h \nabla w^h \cdot u_{\Gamma_\rho} \, dx + \int_\Omega D\phi^h \nabla w^h \cdot \nabla\rho_\phi^h \, dx. \qquad (146)$$

We use the generalized-α method to discretize in time and we follow the numerical formulation described in [56] to avoid singularities and improve the condition number of the linear system. In the simulations, we take $R_0 = 10 \, \mu m$, $D = 10 \, \mu m^2 \, s^{-1}$, $\epsilon = 1 \, \mu m$, $\Gamma_\phi = 0.4 \, \mu m \, s^{-1}$, and $u_{\Gamma_\rho} = (0.6, 0) \, \mu m \, s^{-1}$. The reference problem, defined by Eqs. (136)–(138), is computed using a NURBS mesh that exactly represents the circular domain, which we assume centered at point $(0, 0) \, \mu m$. The mesh is comprised of 100 quadratic elements in the radial direction and 200 in the circumferential direction. The diffuse domain problem is solved in the computational domain $\Omega = [-L, L]$, with $L = 20 \, \mu m$, using 400^2 quadratic NURBS elements.

We solve the sharp-interface and the diffuse-interface problems for two different initial conditions. The first initial condition is $\rho_0(x) = 10 \, H(5 - d(x))$, where H is the Heaviside function. Figure 5(top) shows the isolines of the sharp-interface (ρ^{ref})

---

[15] The variational formulation corresponding to the reference problem is analogous.

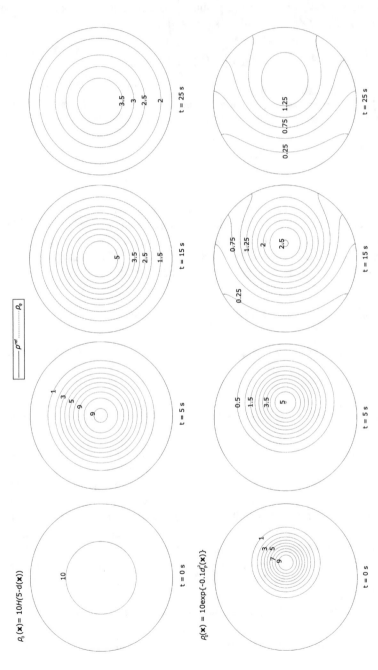

Fig. 5 Time evolution showing the diffusion of the component ρ^{ref} within a circular domain. Comparison between the sharp-interface solution (ρ^{ref}, black solid line) and the diffuse-interface solution (ρ_ϕ, red dashed line) corresponding to two different initial conditions (top and bottom rows, respectively)

and the diffuse-interface (ρ_ϕ) solutions at different times. The solutions obtained are almost identical. For the second initial condition, we take $\rho_0(x) = 10 \exp \big(-0.1\, d_P^2(x) \big)$, where $d_P(x)$ is the distance between x and the point $(2, 0)$ μm. Here, we suppress the symmetry of the problem to further check the accuracy of the phase-field method. Figure 5(bottom) shows the isolines of ρ^{ref} and ρ_ϕ corresponding to the latter initial condition. Again, the results are almost identical. However, in this case a small (almost imperceptible) difference can be observed near the boundary. Better results can be obtained by decreasing the value of ϵ and/or refining the mesh.

References

1. Altundas, Y.B., Caginalp, G.: Computations of dendrites in 3-d and comparison with microgravity experiments. J. Stat. Phys. **110**(3–6), 1055–1067 (2003)
2. Ambrosio, L., Tortorelli, V.M.: Approximation of functional depending on jumps by elliptic functional via Γ-convergence. Commun. Pure Appl. Math. **43**(8), 999–1036 (1990)
3. Anderson, D.M., McFadden, G.B., Wheeler, A.A.: Diffuse-interface methods in fluid mechanics. Annu. Rev. Fluid Mech. **30**(1), 139–165 (1998)
4. Biben, T., Kassner, K., Misbah, C.: Phase-field approach to three-dimensional vesicle dynamics. Phys. Rev. E **72**(4), 041921 (2005)
5. Boettinger, W.J., Warren, J.A., Beckermann, C., Karma, A.: Phase-field simulation of solidification 1. Annu. Rev. Mater. Res. **32**(1), 163–194 (2002)
6. Borden, M.J., Hughes, T.J.R., Landis, C.M., Verhoosel, C.V.: A higher-order phase-field model for brittle fracture: Formulation and analysis within the isogeometric analysis framework. Comput. Methods Appl. Mech. Eng. **273**, 100–118 (2014)
7. Bowen, R.M.: Theory of mixtures. In: Eringen, A.C. (ed.) Continuum Physics, volume III: Mixtures and EM Field Theories, pp. 1–127. Academic Press, New York (1976)
8. Boyer, F., Lapuerta, C., Minjeaud, S., Piar, B., Quintard, M.: Cahn-Hilliard/Navier-Stokes model for the simulation of three-phase flows. Transp. Porous Media **82**(3), 463–483 (2010)
9. Braides, A.: Γ -convergence for Beginners. Oxford University Press, Oxford (2002)
10. Bueno, J., Bona-Casas, C., Bazilevs, Y., Gomez, H.: Interaction of complex fluids and solids: theory, algorithms and application to phase-change-driven implosion. Comput. Mech., 1–14 (2014)
11. Caginalp, G.: Stefan and Hele-Shaw type models as asymptotic limits of the phase-field equations. Phys. Rev. A **39**(11), 5887 (1989)
12. Ceniceros, H.D., Nós, R.L., Roma, A.M.: Three-dimensional, fully adaptive simulations of phase-field fluid models. J. Comput. Phys. **229**(17), 6135–6155 (2010)
13. Chen, L.-Q.: Phase-field models for microstructure evolution. Annu. Rev. Mater. Res. **32**(1), 113–140 (2002)
14. Chen, X., Caginalp, G., Eck, C.: A rapidly converging phase field model. Discrete Contin. Dynam. Syst. **15**(4), 1017 (2006)
15. Coleman, B.D., Noll, W.: The thermodynamics of elastic materials with heat conduction and viscosity. Arch. Ration. Mech. Anal. **13**, 167–178 (1963)
16. Colli, P., Frigeri, S., Grasselli, M.: Global existence of weak solutions to a nonlocal Cahn-Hilliard-Navier-Stokes system. J. Math. Anal. Appl. **386**(1), 428–444 (2012)
17. Cottrell, J.A., Hughes, T.J.R., Bazilevs, Y.: Isogeometric Analysis: Toward Integration of CAD and FEA. Wiley, New York (2009)
18. Dal Maso, G.: An Introduction to Γ-convergence. Springer, New York (1993)
19. Deckelnick, K., Dziuk, G., Elliott, C.M.: Computation of geometric partial differential equations and mean curvature flow. Acta Numer. **14**, 139–232 (2005)

20. Dedeè, L., Borden, M.J., Hughes, T.J.: Isogeometric analysis for topology optimization with a phase field model. Arch. Comput. Methods Eng. **19**(3), 427–465 (2012)
21. Du, Q., Liu, C., Ryham, R., Wang, X.: A phase field formulation of the Willmore problem. Nonlinearity **18**, 1249–1267 (2005)
22. Emmerich, H.: The Diffuse Interface Approach in Materials Science: Thermodynamic Concepts and Applications of Phase-Field Models. Lecture Notes in Physics. Springer, Berlin (2003)
23. Fedkiw, R., Osher, S.: Level Set Methods and Dynamic Implicit Surfaces. Springer, New York (2003)
24. Fife, P.C.: Dynamics of Internal Layers and Diffusive Interfaces, volume 53 of CBMS-NSF Regional Conference Series in Applied Mathematics. Society of Industrial and Applied Mathematics (SIAM), Philadelphia (1988)
25. Gal, C.G., Grasselli, M.: Asymptotic behavior of a Cahn-Hilliard-Navier-Stokes system in 2d. In: Annales de l'Institut Henri Poincare (C) Non Linear Analysis, vol. 27-1, pp. 401–436. Elsevier, New York (2010)
26. Gal, C.G., Grasselli, M.: Instability of two-phase flows: a lower bound on the dimension of the global attractor of the Cahn-Hilliard-Navier-Stokes system. Physica D Nonlinear Phenomena **240**(7), 629–635 (2011)
27. Gomez, H., Hughes, T.J.R.: Provably unconditionally stable, second-order time-accurate, mixed variational methods for phase-field models. J. Comput. Phys. **230**(13), 5310–5327 (2011)
28. Gómez, H., Colominas, I., Navarrina, F., Casteleiro, M.: A discontinuous Galerkin method for a hyperbolic model for convection–diffusion problems in CFD. Int. J. Numer. Methods Eng. **71**(11), 1342–1364 (2007)
29. Gómez, H., Colominas, I., Navarrina, F., Casteleiro, M.: A finite element formulation for a convection–diffusion equation based on Cattaneo's law. Comput. Methods Appl. Mech. Eng. **196**(9), 1757–1766 (2007)
30. Gomez, H., Calo, V.M., Bazilevs, Y., Hughes, T.J.R.: Isogeometric analysis of the Cahn-Hilliard phase-field model. Comput. Methods Appl. Mech. Eng. **197**, 4333–4352 (2008)
31. Gómez, H., Colominas, I., Navarrina, F., Casteleiro, M.: A mathematical model and a numerical model for hyperbolic mass transport in compressible flows. Heat Mass Transfer **45**(2), 219–226 (2008)
32. Gómez, H., Colominas, I., Navarrina, F., París, J., Casteleiro, M.: A hyperbolic theory for advection-diffusion problems: mathematical foundations and numerical modeling. Arch. Comput. Methods Eng. **17**(2), 191–211 (2010)
33. Gomez, H., Reali, A., Sangalli, G.: Accurate, efficient, and (iso) geometrically flexible collocation methods for phase-field models. J. Comput. Phys. **262**, 153–171 (2014)
34. Gonzalez-Ferreiro, B., Gomez, H., Romero, I.: A thermodynamically consistent numerical method for a phase field model of solidification. Commun. Nonlinear Sci. Numer. Simul. **19**(7), 2309–2323 (2014)
35. Gurtin, M.E.: Generalized Ginzburg–Landau and Cahn–Hilliard equations based on a microforce balance. Phys. D **92**, 178–192 (1996)
36. Gurtin, M.E., Polignone, D., Viñals, J.: Two-phase binary fluids and immiscible fluids described by an order parameter. Math. Models Methods Appl. Sci. **6**(6), 815–832 (1996)
37. Gurtin, M.E., Fried, E., Anand, L.: The Mechanics and Thermodynamics of Continua. Cambridge University Press, Cambridge (2009)
38. Haberleitner, M., Jüttler, B., Scott, M.A., Thomas, D.C.: Isogeometric analysis: representation of geometry. In: Encyclopedia of Computational Mechanics Second Edition, pp. 1–24 (2017)
39. Hawkins-Daarud, A., van der Zee, K.G., Oden, J.T.: Numerical simulation of a thermodynamically consistent four-species tumor growth model. Int. J. Numer. Methods Biomed. Eng. **28**, 3–24 (2012)
40. Hou, T.Y., Lowengrub, J.S., Shelley, M.J.: Boundary integral methods for multicomponent fluids and multiphase materials. J. Comput. Phys. **169**(2), 302–362 (2001)

41. Hughes, T.J.R., Sangalli, G.: Mathematics of isogeometric analysis: a conspectus. In: Encyclopedia of Computational Mechanics Second Edition, pp. 1–40 (2018)
42. Hughes, T.J.R., Cottrell, J.A., Bazilevs, Y.: Isogeometric analysis: CAD, finite elements, NURBS, exact geometry and mesh refinement. Comput. Methods Appl. Mech. Eng. **194**, 4135–4195 (2005)
43. Jacqmin, D.: Calculation of two-phase Navier-Stokes flows using phase-field modeling. J. Comput. Phys. **155**(1), 96–127 (1999)
44. Jansen, K.E., Whiting, C.H., Hulbert, G.M.: A generalized-α method for integrating the filtered Navier-Stokes equations with a stabilized finite element method. Comput. Methods Appl. Mech. Eng. **190**, 305–319 (2000)
45. Kay, D., Styles, V., Welford, R.: Finite element approximation of a Cahn-Hilliard-Navier-Stokes system. Interfaces Free Bound **10**(1), 15–43 (2008)
46. Kim, J., Lowengrub, J.: Phase field modeling and simulation of three-phase flows. Interfaces Free Bound. **7**, 435–466 (2005)
47. Kim, J., Kang, K., Lowengrub, J.: Conservative multigrid methods for Cahn-Hilliard fluids. J. Comput. Phys. **193**(2), 511–543 (2004)
48. Kobayashi, R.: A numerical approach to three-dimensional dendritic solidification. Exp. Math. **3**(1), 59–81 (1994)
49. Liu, C., Shen, J.: A phase field model for the mixture of two incompressible fluids and its approximation by a Fourier-spectral method. Physica D Nonlinear Phenomena **179**(3), 211–228 (2003)
50. Liu, J., Gomez, H., Evans, J.A., Hughes, T.J.R., Landis, C.M.: Functional entropy variables: a new methodology for deriving thermodynamically consistent algorithms for complex fluids, with particular reference to the isothermal Navier-Stokes-Korteweg equations. J. Comput. Phys. **248**, 47–86 (2013)
51. Liu, J., Landis, C.M., Gomez, H., Hughes, T.J.R.: Liquid-vapor phase transition: thermo-mechanical theory, entropy stable numerical formulation, and boiling simulations. Comput. Methods Appl. Mech. Eng. **297**, 476–553 (2015)
52. Lorenzo, G., Scott, M.A., Tew, K.B., Hughes, T.J.R., Zhang, Y.J., Liu, L., Vilanova, G., Gomez, H.: Tissue-scale, personalized modeling and simulation of prostate cancer growth. Proc. Natl. Acad. Sci. **113**(48), E7663–E7671 (2016)
53. Lowengrub, J., Truskinovsky, L.: Quasi-incompressible Cahn-Hilliard fluids and topological transitions. Proc. R. Soc. Lond. A Math. Phys. Eng. Sci. **454**(1978), 2617–2654 (1998)
54. Lowengrub, J.S., Rätz, A., Voigt, A.: Phase-field modeling of the dynamics of multicomponent vesicles: Spinodal decomposition, coarsening, budding, and fission. Phys. Rev. E **79**, 031926 (2009)
55. Miehe, C., Welschinger, F., Hofacker, M.: Thermodynamically consistent phase-field models of fracture: Variational principles and multi-field FE implementations. Int. J. Numer. Methods Eng. **83**(10), 1273–1311 (2010)
56. Moure, A., Gomez, H.: Phase-field model of cellular migration: three-dimensional simulations in fibrous networks. Comput. Methods Appl. Mech. Eng. **320**, 162–197 (2017)
57. Oden, J.T., Hawkins, A., Prudhomme, S.: General diffuse-interface theories and an approach to predictive tumor growth modeling. Math. Models Methods Appl. Sci. **20**(03), 477–517 (2010)
58. Penrose, O., Fife, P.C.: Thermodynamically consistent models of phase-field type for the kinetic of phase transitions. Phys. D **43**(1), 44–62 (1990)
59. Provatas, N., Elder, K.: Phase-Field Methods in Materials Science and Engineering. Wiley-VCH, Weinheim (2010)
60. Queutey, P., Visonneau, M.: An interface capturing method for free-surface hydrodynamic flows. Comput. Fluids **36**(9), 1481–1510 (2007)
61. Romano, A., Marasco, A.: Continuum Mechanics: Advanced Topics and Research Trends. Modeling and Simulation in Science, Engineering and Technology. Springer, New York (2010)
62. Shao, D., Levine, H., Rappel, W.-J.: Coupling actin flow, adhesion, and morphology in a computational cell motility model. Proc. Natl. Acad. Sci. **109**(18), 6851–6856 (2012)

63. Truesdell, C., Noll, W.: The Non-Linear Field Theories of Mechanics. Springer, New York (1965)
64. Vilanova, G., Gomez, H., Colominas, I.: A numerical study based on the FEM of a multiscale continuum model for tumor angiogenesis. J. Biomech. **45**, S466 (2012)
65. Vilanova, G., Colominas, I., Gomez, H.: Capillary networks in tumor angiogenesis: from discrete endothelial cells to phase-field averaged descriptions via isogeometric analysis. Int. J. Numer. Methods Biomed. Eng. **29**(10), 1015–1037 (2013)
66. Vilanova, G., Colominas, I., Gomez, H.: Coupling of discrete random walks and continuous modeling for three-dimensional tumor-induced angiogenesis. Comput. Mech. **53**(3), 449–464 (2014)
67. Wang, S.-L., Sekerka, R.F., Wheeler, A.A., Murray, B.T., Coriel, S.R., Braun, R.J., McFadden, G.B.: Thermodynamically-consistent phase-field models for solidification. Phys. D **69**, 189–200 (1993)
68. Xu, J., Vilanova, G., Gomez, H.: Phase-field model of vascular tumor growth: Three-dimensional geometry of the vascular network and integration with imaging data. Comput. Methods Appl. Mech. Eng. **359**, 112648 (2020)

Consistent Internal Energy Based Schemes for the Compressible Euler Equations

T. Gallouët, R. Herbin, J.-C. Latché, and N. Therme

Abstract Numerical schemes for the solution of the Euler equations have recently been developed, which involve the discretisation of the internal energy equation, with corrective terms to ensure the correct capture of shocks, and, more generally, the consistency in the Lax-Wendroff sense. These schemes may be staggered or colocated, using either structured meshes or general simplicial or tetrahedral/hexahedral meshes. The time discretization is performed by fractional-step algorithms; these may be either based on semi-implicit pressure correction techniques or segregated in such a way that only explicit steps are involved (referred to hereafter as "explicit" variants). In order to ensure the positivity of the density, the internal energy and the pressure, the discrete convection operators for the mass and internal energy balance equations are carefully designed; they use an upwind technique with respect to the material velocity only. The construction of the fluxes thus does not need any Riemann or approximate Riemann solver, and yields easily implementable algorithms. The stability is obtained without restriction on the time step for the pressure correction scheme and under a CFL-like condition for explicit variants: preservation of the integral of the total energy over the computational domain, and positivity of the density and the internal energy. The semi-implicit first-order upwind scheme satisfies a local discrete entropy inequality. If a MUSCL-like scheme is used in order to limit the scheme diffusion, then a weaker property holds: the entropy inequality is satisfied up to a remainder term which is shown to tend to

T. Gallouët · R. Herbin (✉)
Aix-Marseille Université, Marseille, France
e-mail: thierry.gallouet@univ-amu.fr; raphaele.herbin@univ-amu.fr

J.-C. Latché
Institut de Radioprotection et de Sûreté Nucléaire (IRSN), Paris, France
e-mail: jean-claude.latche@irsn.fr

N. Therme
CEA/CESTA, Le Barp, France
e-mail: nicolas.therme@cea.fr

© The Author(s), under exclusive license to Springer Nature Switzerland AG 2021
D. Greiner et al. (eds.), *Numerical Simulation in Physics and Engineering: Trends and Applications*, SEMA SIMAI Springer Series 24,
https://doi.org/10.1007/978-3-030-62543-6_3

zero with the space and time steps, if the discrete solution is controlled in L^∞ and BV norms. The explicit upwind variant also satisfies such a weaker property, at the price of an estimate for the velocity which could be derived from the introduction of a new stabilization term in the momentum balance. Still for the explicit scheme, with the above-mentioned MUSCL-like scheme, the same result only holds if the ratio of the time to the space step tends to zero.

Keywords Compressible flows · Euler equations · Internal energy · Pressure correction · Segregated algorithms · Entropy estimates

1 Introduction

We address in this paper the solution of the Euler equations for an ideal gas, which read:

$$\partial_t \rho + \text{div}(\rho \, \boldsymbol{u}) = 0, \tag{1a}$$

$$\partial_t (\rho \, \boldsymbol{u}) + \text{div}(\rho \, \boldsymbol{u} \otimes \boldsymbol{u}) + \nabla p = 0, \tag{1b}$$

$$\partial_t (\rho \, E) + \text{div}(\rho \, E \, \boldsymbol{u}) + \text{div}(p \, \boldsymbol{u}) = 0, \tag{1c}$$

$$p = (\gamma - 1) \, \rho \, e, \qquad E = \frac{1}{2}|\boldsymbol{u}|^2 + e, \tag{1d}$$

where t stands for the time, ρ, $\boldsymbol{u}$, p, E and e are the density, velocity, pressure, total energy and internal energy respectively, and $\gamma > 1$ is a coefficient specific to the considered fluid. The problem is supposed to be posed over $\Omega \times (0, T)$, where Ω is an open bounded connected subset of $\mathbb{R}^d$, $1 \le d \le 3$, and $(0, T)$ is a finite time interval. System (1) is complemented by initial conditions for ρ, e and $\boldsymbol{u}$, let us say ρ_0, e_0 and $\boldsymbol{u}_0$ respectively, with $\rho_0 > 0$ and $e_0 > 0$, and by suitable boundary conditions which we suppose to be $\boldsymbol{u} \cdot \boldsymbol{n} = 0$ at any time and a.e. on $\partial\Omega$, where $\boldsymbol{n}$ stands for the normal vector to the boundary.

Finite volume schemes for the solution of hyperbolic problems such as the system (1) generally use a collocated arrangement of the unknowns, which are associated to the cell centers, and apply a Godunov-like technique for the computation of the fluxes at the cells faces: the face is seen as a discontinuity line for the beginning-of-time-step numerical solution, supposed to be constant in the two adjacent cells; the value of the solution of the so-posed Riemann problem on the discontinuity line is computed, either exactly or approximately; the numerical solution at the end-of-time-step is computed with these values, and is a piecewise constant function (see e.g. [3, 39] for the development of such solvers). In one space dimension, this method consists, at least for exact Riemann solvers, in a projection of the exact solution. Then, thanks to the properties of the projection, this process applied to the Euler equations yields consistent schemes which preserve the non-negativity of the density and the internal energy and, for first-order variants, satisfy an entropy inequality. The price to pay is the computational cost of the evaluation of the fluxes, and the fact that this issue is intricate enough to put almost out of reach implicit-in-time formulations, which would allow to relax CFL time step constraints. In addition, preserving the accuracy for low Mach number flows is a difficult task (see e.g. [18] and references herein).

The aim here is first to review some recent schemes which follow a different route, and then prove some discrete entropy estimates and/or consistency results for these schemes. The space discretization may be colocated [25] or staggered [17, 21, 22]: in the colocated case, all unknowns are located at the center of the discretization cells, while in the staggered case, scalar variables are associated to cell centers while the velocity is associated to the faces, or, equivalently, to staggered mesh(es). The use of staggered discretization for compressible flows goes back to the MAC scheme [19], and has been the subject of a wide litterature (see [42] for a textbook and references in [17, 21, 22]). Staggered discretizations have been preferred in the open source CALIF^{3}S [4] used for nuclear safety applications because the resulting semi-implicit schemes are asymptotically stable in the low Mach number regime [24]. Two different staggered space discretizations may be considered: either the so-called Marker-And-Cell (MAC) scheme for structured grids [20] or, for general meshes, a space discretization using degrees of freedom similar to low-order Rannacher-Turek [34] or Crouzeix-Raviart [8] finite elements (see Fig. 1). With this space discretization, the use of Riemann solvers seems difficult (scalar unknowns and velocities may still be considered as piecewise constant functions, but not associated to the same partition of the computational domain). The positivity of the internal energy is thus ensured by a non-standard argument: the internal energy balance is discretized instead of the actual (total) energy balance (1c) by a positivity-preserving scheme. This strategy is known to lead to consistency problems (wrong shock speeds for instance), which are

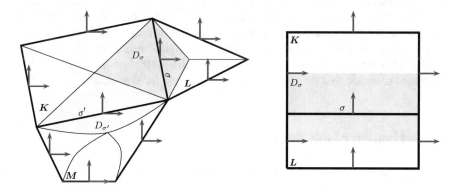

Fig. 1 Meshes and unknowns—Left: unstructured discretizations (the present sketch illustrates the possibility, implemented in our software CALIF^{3}S [4], of mixing simplicial and quadrangular cells); scalars variables are associated to the primal cells (here K, L and M) while velocity vectors are associated to the faces (here, σ and σ') or, equivalently, to dual cells (here, D_σ and $D_{\sigma'}$)— Right: MAC discretization; scalars variables are associated to the primal cells and each face is associated to the component of the velocity normal to the face

circumvented by some correction terms in the discrete internal energy correction. Until now, the use of the internal energy equation associated to a consistency correction seems to be restricted to the context of Lagrangian approaches, up to a very recent work implementing a Lagrange-remap technique on staggered meshes [9], and some recent developments extending the techniques developed here to more general meshes [31]. Two time discretizations are proposed: a pressure correction technique and a segregated scheme involving only explicit steps. The resulting schemes offer many interesting properties: both the density and internal energy positivity are preserved, unconditionally for the pressure correction scheme and under CFL-like conditions for the segregated explicit variant, and the integral of the total energy on the computational domain is conserved (which yields a stability result); the construction of the fluxes simply relies on standard upwinding techniques of the convection operators with respect to the material velocity; finally, the space approximation, the fluxes and the choice of the internal energy balance are consistent with usual discretizations of quasi-incompressible flows, so the pressure correction scheme is asymptotic preserving by construction in the limit of vanishing low Mach number flows (see [24] for a study in the case of the barotropic Euler equations).

In addition, a discrete entropy estimate is obtained for the (upwind) pressure correction scheme, while only a conditional weak entropy estimate seems to hold for the segregated explicit variant. Note that the schemes studied here belong to a class often referred to as "flux splitting schemes" in the literature, since they may be obtained by splitting the system by a two-step technique (usually into a "convective" and "acoustic" part), applying a standard scheme to each part (which, for the convection system, indeed yields, at first order, an upwinding with respect to the material velocity) and then summing both steps to obtain the final flux. Works in this direction may be found in [29, 30, 37, 40, 43], and we hope that the discussion presented on the entropy may be extended in some way to these numerical methods.

The paper is organised as follows; Sect. 2 is devoted to the derivation of the previously mentioned internal energy based schemes in the semi-discrete time setting. Section 3 presents some new and original results concerning some entropy estimates and/or entropy consistency which hold for both the colocated and staggered schemes, first for implicit schemes, and then for explicit schemes.

2   Derivation of the Numerical Schemes

2.1   A Basic Result on Convection Operators

Let ρ and $\boldsymbol{u}$ be regular respectively scalar and vector-valued functions such that

$$\partial_t \rho + \text{div}(\rho \boldsymbol{u}) = 0.$$

Let z be a regular scalar function. Then

$$\mathcal{C}(z) = \partial_t(\rho z) + \text{div}(\rho z \boldsymbol{u}) = \rho(\partial_t z + \boldsymbol{u} \cdot \nabla z) + z(\partial_t \rho + \text{div}(\rho \boldsymbol{u}))$$
$$= \rho(\partial_t z + \boldsymbol{u} \cdot \nabla z). \tag{2}$$

Let φ be a regular real function. Then:

$$\varphi'(z)\,\mathcal{C}(z) = \varphi'(z)\,\rho\left(\partial_t z + \boldsymbol{u} \cdot \nabla z\right) = \rho\left(\partial_t \varphi(z) + \boldsymbol{u} \cdot \nabla \varphi(z)\right).$$

Now, reversing the computation performed in Relation (2) with $\varphi(z)$ instead of z leads to

$$\varphi'(z)\,\mathcal{C}(z) = \partial_t\left(\rho\varphi(z)\right) + \text{div}\left(\rho\varphi(z)\,\boldsymbol{u}\right). \tag{3}$$

The following lemma states a time semi-discrete version of this computation.

Lemma 1 (Convection Operator) *Let ρ^n, ρ^{n+1}, z^n and z^{n+1} be regular scalar functions, let $\boldsymbol{u}$ be a regular vector-valued function and let φ be a twice-differentiable real function. Let us suppose that*

$$\frac{1}{\delta t}\,(\rho^{n+1} - \rho^n) + \text{div}(\rho^{n+1}\boldsymbol{u}) = 0, \tag{4}$$

with δt a positive real number. Then

$$\varphi'(z^{n+1})\left[\frac{1}{\delta t}\,(\rho^{n+1}z^{n+1} - \rho^n z^n) + \text{div}(\rho^{n+1}z^{n+1}\boldsymbol{u})\right]$$
$$= \frac{1}{\delta t}\left(\rho^{n+1}\varphi(z^{n+1}) - \rho^n\varphi(z^n)\right) + \text{div}\left(\rho^{n+1}\varphi(z^{n+1})\,\boldsymbol{u}\right) + \mathcal{R}^n, \tag{5}$$

with

$$\mathcal{R}^n = \frac{1}{2\,\delta t}\rho^n\varphi''(\bar{z})\,(z^{n+1} - z^n)^2, \quad \bar{z} = \theta z^n + (1 - \theta)z^{n+1}, \quad \theta \in [0, 1].$$

Proof We first begin by deriving a discrete analogue to Identity (2):

$$\frac{1}{\delta t}\,(\rho^{n+1}z^{n+1} - \rho^n z^n) + \text{div}(\rho^{n+1}z^{n+1}\boldsymbol{u})$$
$$= \frac{1}{\delta t}\,\rho^n\,(z^{n+1} - z^n) + \rho^{n+1}\boldsymbol{u} \cdot \nabla z^{n+1} + z^{n+1}\left[\frac{1}{\delta t}\,(\rho^{n+1} - \rho^n) + \text{div}(\rho^{n+1}\boldsymbol{u})\right]$$
$$= \frac{1}{\delta t}\,\rho^n\,(z^{n+1} - z^n) + \rho^{n+1}\boldsymbol{u} \cdot \nabla z^{n+1}. \tag{6}$$

Then the result follows by multiplying this relation by $\varphi'(z^{n+1})$, using a Taylor expansion for the first term and the same combination of partial derivative as in the continuous case for the second term, and finally, still as in the continuous case, by performing this computation in the reverse sense with $\varphi(z^n)$ and $\varphi(z^{n+1})$ instead of z^n and z^{n+1}.

2.2   Internal Energy Formulation

We begin with a formal reformulation of the energy equation. Let us suppose that the solution is regular, and let E_k be the kinetic energy, defined by $E_k = \frac{1}{2}|u|^2$. Taking the inner product of (1b) by u yields, after the formal compositions of partial derivatives described in the previous section:

$$\partial_t(\rho E_k) + \text{div}(\rho E_k u) + \nabla p \cdot u = 0. \tag{7}$$

This relation is referred to as the kinetic energy balance. Subtracting this relation to the total energy balance (1c), we obtain the so-called internal energy balance equation:

$$\partial_t(\rho e) + \text{div}(\rho e u) + p\,\text{div} u = 0. \tag{8}$$

Since,

- as seen in the previous section, thanks to the mass balance equation, the first two terms in the left-hand side of (8) may be recast as a transport operator,
- and, from the equation of state, the pressure vanishes when $e = 0$,

this equation implies that, if $e \geq 0$ at $t = 0$ and with suitable boundary conditions, then e remains non-negative at all time. The same result would hold if (8) featured a non-negative right-hand side, as for the compressible Navier-Stokes equations. Solving the internal energy balance (8) instead of the total energy balance (1c) is thus appealing, to preserve this positivity property by construction of the scheme. In addition, it avoids introducing a space discretization for the total energy which, for a staggered discretization, combines cell-centered (internal energy and density) and face-centered (velocity) variables. However, a raw discretization of a non-conservative equation derived from a conservative system (formally, i.e. supposing unrealistic regularity properties of the solution) may be non-consistent (and the numerical test presented in Sect. 2.6 shows that, for the problem at hand, such a scheme would be unable to capture shock solutions). To deal with this problem, we implement the following strategy:

- First, we derive a discrete kinetic energy balance, by mimicking at the discrete level the computation leading to Eq. (7), so as to identify the terms which are likely to lead to non-consistency: the numerical diffusion in the momentum balance equation yields dissipation terms in the kinetic energy balance which are observed to behave, when the space and time step tend to zero, as measure born by the shocks which modify the jump conditions.

– These terms are thus compensated in the internal energy balance.

At the fully discrete level, for staggered discretizations, the kinetic and internal energy balances are not posed on the same mesh (the dual and primal mesh respectively); however, it is possible to derive from the kinetic energy balance on the dual mesh a counterpart posed on the primal mesh and, adding to the internal energy balance yields a conservative total energy balance. The scheme then can be proven to be consistent in the Lax-Wendroff sense to the weak form of the total energy balance: for a given sequence of discrete solutions (obtained with a sequence of discretizations whith space and time steps tending to zero) controlled and converging to a limit in suitable norms (namely, uniformly bounded and converging in L^r norms, for $r \in [1, +\infty)$), we show that the limit is a weak solution to the Euler equations [22, 23]. In the colocated case, both kinetic and internal energy balances are posed on the same mesh and a discrete local total energy balance is easily recovered [25].

2.3   The Time Semi-discrete Pressure Correction Scheme

This semi-discrete pressure correction scheme takes the following general form:

$$\frac{1}{\delt}(\rho^n \, \tilde{u}^{n+1} - \rho^{n-1} \, u^n) + \operatorname{div}(\rho^n \, u^n \otimes \tilde{u}^{n+1}) + \zeta^n \nabla p^n = 0, \tag{9a}$$

$$\frac{1}{\delt}\rho^n (u^{n+1} - \tilde{u}^{n+1}) + \nabla p^{n+1} - \zeta^n \nabla p^n = 0, \tag{9b}$$

$$\frac{1}{\delt}(\rho^{n+1} - \rho^n) + \operatorname{div}(\rho^{n+1} \, u^{n+1}) = 0, \tag{9c}$$

$$\frac{1}{\delt}(\rho^{n+1} \, e^{n+1} - \rho^n \, e^n) + \operatorname{div}(\rho^{n+1} \, e^{n+1} \, u^{n+1}) + p^{n+1} \operatorname{div} u^{n+1} = S^{n+1}, \tag{9d}$$

$$p^{n+1} = (\gamma - 1) \, \rho^{n+1} \, e^{n+1}. \tag{9e}$$

Solving the first equation yields a tentative velocity $\tilde{u}^{n+1}$; this is the velocity prediction step, which is decoupled from the other equations of the system. Equations (9b)–(9e) constitute the correction step and are solved simultaneously; in the relation (9d), the term $\rho^{n+1} e^{n+1}$ is recast as a function of the pressure only thanks to the equation of state (1d) and the velocity u^{n+1} is eliminated thanks to the divergence of (9b) divided by ρ^n. The result is a nonlinear and nonconservative elliptic problem for the pressure only. This process must be performed at the fully discrete level to preserve the properties of the scheme. The coefficient ζ^n in Eq. (9a) and the correction term S^{n+1} in (9d) are chosen so as to ensure stability and consistency, as shown below. The first step of this process is to obtain a discrete kinetic energy balance. To this purpose, let us multiply (9a) by $\tilde{u}^{n+1}$ and apply Lemma 1 component by component, with $\varphi(s) = \frac{1}{2}s^2$. We get:

$$\frac{1}{2\,\delta t}\left(\rho^n\,|\tilde{u}^{n+1}|^2 - \rho^{n-1}\,|u^n|^2\right) + \frac{1}{2}\mathrm{div}\left(\rho^n\,|\tilde{u}^{n+1}|^2 u^n\right) + \zeta^n \nabla p^n \cdot \tilde{u}^{n+1} + R_1^n = 0,$$

(10)

with

$$R_1^n = \frac{1}{2\,\delta t}|\tilde{u}^{n+1} - u^n|^2.$$

Note that the mass balance equation (9c), which is a fundamental assumption in Lemma 1, only holds at this stage of the algorithm with the previous time step values, hence the shift of the time level of the density in (9a). Let us now recast Eq. (9b) as

$$\alpha^n u^{n+1} + \frac{1}{\alpha^n}\nabla p^{n+1} = \alpha^n \tilde{u}^{n+1} + \frac{\zeta^n}{\alpha^n}\nabla p^n, \quad \alpha^n = \left[\frac{\rho^n}{\delta t}\right]^{1/2}$$

and square this relation, to get

$$\frac{1}{2\,\delta t}\rho^n\,|u^{n+1}|^2 + \nabla p^{n+1}\cdot u^{n+1} + R_2^n = \frac{1}{2\,\delta t}\rho^n\,|\tilde{u}^{n+1}|^2 + \zeta^n\nabla p^n\cdot\tilde{u}^{n+1},$$

(11)

with

$$R_2^n = \frac{\delta t}{\rho^n}|\nabla p^{n+1}|^2 - (\zeta^n)^2\frac{\delta t}{\rho^n}|\nabla p^n|^2.$$

Summing (10) and (11) yields the kinetic energy balance that we are seeking:

$$\frac{1}{2\,\delta t}\left(\rho^n\,|u^{n+1}|^2 - \rho^{n-1}\,|u^n|^2\right) + \frac{1}{2}\mathrm{div}\left(\rho^n\,|\tilde{u}^{n+1}|^2 u^n\right) + \nabla p^{n+1}\cdot u^{n+1} + R_1^n + R_2^n = 0.$$

The coefficient ζ^n is then chosen in such a way that the remainder term R_2^n is a difference of two consecutive time levels of the same quantity; this is the case for

$$\zeta^n = \left[\frac{\rho^n}{\rho^{n-1}}\right]^{1/2}.$$

Supposing the control in $L^1(0, T, BV)$ of the pressure and in L^∞ of the pressure and of the inverse of the density, the term R_2^n may thus be seen to tend with zero with the discretization parameters in a distributional sense. The term R_1^n is compensated in the internal energy balance, by choosing $S^{n+1} = R_1^n$, thus ensuring that $S^{n+1} \geq 0$. The definition of the time-discrete scheme is now complete.

2.4 The Fully Discrete Pressure Correction Scheme

The fully discrete scheme is obtained from System (9) by applying the following guidelines:

- The mass and internal energy balances (i.e. Eqs. (9c) and (9d) respectively) are discretized on the primal mesh, while the velocity prediction (9a) and correction (9b) are discretized on the dual mesh(es). The equation of state only involves cell quantities, and its expression is obtained by writing (9e) for these latter.
- The space arrangement of the unknowns (density discretized at the cell and velocity at the faces) yields a natural expression of the mass fluxes in the mass balance, performed by a first-order upwind scheme (with respect to the velocity). By construction, the density is thus non-negative; in fact at the discrete level, it remains positive if the initial density is positive. The discrete mass balance equation on the cell K whose measure is denoted by $|K|$ takes the form:

$$\frac{|K|}{\delta t}(\rho_K^{n+1} - \rho_K^n) + \sum_{\sigma \in \mathcal{E}(K)} F_{K,\sigma}^{n+1} = 0, \tag{12}$$

where $\mathcal{E}(K)$ denotes the set of edges of K and $F_{K,\sigma}$ is the mass flux across σ outward K.
- Let $\mathcal{C}_K(e^{n+1})$ denote the sum of the discrete time-derivative and convection operator in the internal energy balance (9d); this quantity reads:

$$\mathcal{C}_K(e^{n+1}) = \frac{|K|}{\delta t}(\rho_K^{n+1}e_K^{n+1} - \rho_K^n e_K^n) + \sum_{\sigma \in \mathcal{E}(K)} F_{K,\sigma}e_\sigma^{n+1},$$

where e_σ^{n+1} is the upwind approximation of e^{n+1} at σ with respect to $F_{K,\sigma}^{n+1}$ (or, equivalently, since the density is positive, with respect to the velocity). The structure of $\mathcal{C}_K(e^{n+1})$ (precisely speaking, the fact that $\mathcal{C}_K(e^{n+1})$ vanishes thanks to the mass balance if the internal energy e^{n+1} is constant over Ω) was shown in [27] to yield a positivity-preserving operator, and is also a necessary condition for a fully discrete version of Lemma 1 to hold; this is of course linked since both results rely on the possibility to recast $\mathcal{C}_K$ as a transport operator, and the positivity-preserving property of $\mathcal{C}_K$ may be proved by applying Lemma 1 with $\varphi(s) = \min(s, 0)^2$. Once again, thanks to the arrangement of the unknowns, a natural discretization for $\mathrm{div}u^{n+1}$ is available. Since p^{n+1} is a function of e^{n+1} (given by the equation of state) which vanishes for $e^{n+1} = 0$ and since the corrective term is non-negative, we are able to show that the discrete internal energy is kept positive by the scheme.

- To allow to derive a discrete kinetic energy balance, the same structure is needed for the time-derivative and convection operator in the velocity prediction step (9a). This raises a difficulty since this equation is posed on the dual mesh, and thus we need an analogue of the mass balance (12) to also hold on this mesh. The way to build the face density and the mass fluxes across the faces of the dual mesh for such a relation to hold, while still ensuring the scheme consistency, is a central ingredient of the scheme; it is detailed in [13] for the MAC discretization and in [28] for unstructured discretizations.

Once the face density is defined, the discretization of the coefficient ζ^n is straightforward. In order to combine the discrete equivalents of $\boldsymbol{u} \cdot \boldsymbol{\nabla} p$ (kinetic energy balance) and $p \operatorname{div} \boldsymbol{u}$ (internal energy balance), the discrete gradient is defined as the transposed of the divergence operator with respect to the L^2 inner product (if $\boldsymbol{u} \cdot \boldsymbol{\nabla} p + p \operatorname{div} \boldsymbol{u} = \operatorname{div}(p \, \boldsymbol{u})$, the integral of this quantity over the computational domain vanishes when the normal velocity is prescribed to zero at the boundary). Note that this definition is consistent with the usual treatment in the incompressible case, and is a key ingredient for the scheme to be asymptotic preserving in the limit of vanishing Mach number flows [24]. As in the incompressible case, it also allows to control the L^2 norm of the pressure by a weak norm of its gradient, which is central for convergence studies; with this respect, a discrete *inf-sup* condition is required in some sense, which is true for staggered discretizations.

2.5 A Segregated Variant

A variant of the proposed scheme which consists only in explicit steps (in the sense that these steps do not require the solution of any linear or non-linear algebraic system) reads, in the time semi-discrete setting:

$$\frac{1}{\delta t}(\rho^{n+1} - \rho^n) + \operatorname{div}(\rho^n \, \boldsymbol{u}^n) = 0, \tag{13a}$$

$$\frac{1}{\delta t}(\rho^{n+1} e^{n+1} - \rho^n e^n) + \operatorname{div}(\rho^n e^n \, \boldsymbol{u}^n) + p^n \operatorname{div} \boldsymbol{u}^n = S^n, \tag{13b}$$

$$p^{n+1} = (\gamma - 1) \rho^{n+1} e^{n+1}, \tag{13c}$$

$$\frac{1}{\delta t}(\rho^{n+1} \boldsymbol{u}^{n+1} - \rho^n \, \boldsymbol{u}^n) + \operatorname{div}(\rho^n \, \boldsymbol{u}^n \otimes \boldsymbol{u}^n) + \boldsymbol{\nabla} p^{n+1} = 0. \tag{13d}$$

The update of the pressure before the solution of the momentum balance equation is crucial in our derivation of entropy estimates (see Sect. 3 below). This issue seems to be supported by numerical experiments: omitting it, we observe the appearance of non-entropic discontinuities in rarefaction waves [22].

The space discretization differs from the pressure correction scheme described in the above section in two points:

- the discretization of the convection operator in the momentum balance equation (13d) is performed by the first order upwind scheme (still with respect to the material velocity u^n),
- the corrective term S^n is still obtained by deriving a kinetic energy balance multiplying Eq. (13d) by u^{n+1}, but its expression is quite different, due to the time-level used in the convection operator. The time-discretization is now anti-diffusive but, as usual for explicit schemes, this anti-diffusion is counterbalanced by the diffusion in the approximation of the convection (hence the upwinding) and S^n is non-negative only under a CFL condition.

2.6 A Numerical Test

In this section, we reproduce a test performed in [21] to assess the behaviour of the scheme on a one dimensional Riemann problem. We choose initial conditions such that the structure of the solution consists in two shock waves, separated by the contact discontinuity, with sufficiently strong shocks to allow an easy discrimination of correct numerical solutions. These initial conditions are those proposed in [39, chapter 4], for the test referred to as Test 5. The computations are performed with the open-source software CALIF^{3}S [4].

The density fields obtained with $h = 1/2000$ (or a number of cells $n = 2000$) at $t = 0.035$, with and without assembling the corrective source term in the internal energy balance, together with the analytical solution, are shown on Fig. 2. We observe that both schemes seem to converge, but the corrective term is necessary to obtain the right solution. Without a corrective term, one can check that the obtained solution is not a weak solution to the Euler system (Rankine-Hugoniot conditions are not verified). We also observe that the scheme is rather diffusive especially at contact discontinuities for which the beneficial compressive effect of the shocks does not apply; this may be cured in the explicit variant by implementing MUSCL-like algorithms [14].

Extensive multidimensional tests were performed in both the staggered case [17] and the colocated case [25].

3 Entropy

In the case of regular solutions to the Euler equations (1), an additional conservation law can be written for an additional quantity called entropy; however, in the presence of shock waves, the (mathematical) entropy decreases. It is now known that weak

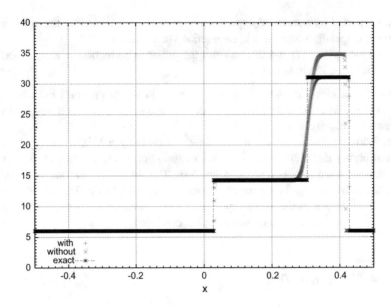

Fig. 2 Test 5 of [39, chapter 4]—density obtained with $n = 2000$ cells, with and without corrective source terms in pressure correction scheme, and analytical solution

solutions of the Euler system satisfying an entropy inequality may be non unique [5]; nevertheless, entropy inequalities play an important role in providing global stability estimates [3].

When solving the Euler equations numerically, it is thus natural to design numerical schemes such that some entropy inequalities are satisfied by the approximate solutions; these inequalities should enable to prove that, as the mesh and time steps tend to 0, the limit of the approximate solutions, if it exists, satisfies an entropy inequality. A classical way of doing so is to design so-called "entropy stable schemes" [38]. Discrete entropy inequalities are known for the one dimensional case for the Godunov scheme [15] and have been derived for Roe-type schemes in the one space dimension case [26]. Entropy stability has also been proven in the multidimensional case for semi-discrete schemes on unstructured meshes [32, 35]. In the sequel we show that an implicit upwind scheme (at least with the upwinding with respect to the material velocity used here) is indeed entropy stable. However it is not always possible to obtain entropy stability, especially for fully discrete schemes such as the explicit schemes studied below; in this case, weaker discrete entropy inequalities or estimates are obtained which allow to fulfil our goal, namely to show that the possible limits of the approximate solutions satisfy an entropy inequality. Such a technique was used for the convergence study of a time implicit mixed finite volume–finite element scheme for the Euler-Fourier equations [11], with a special equation of state which allows to obtain a priori estimates.

Both the pressure correction scheme (9) and the segregated scheme (13) involve a discrete equivalent of the following subsystem:

$$\partial_t \rho + \mathrm{div}(\rho\, \boldsymbol{u}) = 0, \tag{14a}$$

$$\partial_t (\rho\, e) + \mathrm{div}(\rho\, e\, \boldsymbol{u}) + p\, \mathrm{div}(\boldsymbol{u}) = \mathcal{R} \geq 0, \tag{14b}$$

$$p = (\gamma - 1)\, \rho\, e, \tag{14c}$$

with the same initial and boundary conditions as for the full system (1).

The derivation of an entropy for the continuous Euler system may be deduced from the subsystem (14) in the following way. We seek an entropy function η satisfying:

$$\partial_t \eta(\rho, e) + \mathrm{div}\big[\eta(\rho, e)\, \boldsymbol{u}\big] \leq 0. \tag{15}$$

To this end, we introduce the functions φ_ρ and φ_e defined as follows:

$$\varphi_\rho(z) = z \ln(z), \quad \varphi_e(z) = \frac{-1}{\gamma - 1} \ln(z), \quad \text{for } z > 0. \tag{16}$$

For regular functions, the function η defined by

$$\eta(\rho, e) = \varphi_\rho(\rho) + \rho \varphi_e(e) \tag{17}$$

satisfies (15). Indeed, multiplying (14a) by $\varphi_\rho'(\rho)$, a formal computation yields:

$$\partial_t \big[\varphi_\rho(\rho)\big] + \mathrm{div}\big[\varphi_\rho(\rho)\, \boldsymbol{u}\big] + \big[\rho \varphi_\rho'(\rho) - \varphi_\rho(\rho)\big]\mathrm{div}(\boldsymbol{u}) = 0. \tag{18}$$

Then, multiplying (14b) by $\varphi_e'(e)$ yields, once again formally, since $\varphi_e'(z) < 0$ for $z > 0$:

$$\partial_t \big[\rho\, \varphi_e(e)\big] + \mathrm{div}\big[\rho\, \varphi_e(e)\, \boldsymbol{u}\big] + \varphi_e'(e)\, p\, \mathrm{div}(\boldsymbol{u}) \leq 0. \tag{19}$$

Summing (18) and (19) and noting that φ_ρ and φ_e are chosen such that

$$\rho \varphi_\rho'(\rho) - \varphi_\rho(\rho) + \varphi_e'(e)\, p = 0, \tag{20}$$

we obtain (15), which is an entropy balance for the Euler equations, for the specific entropy defined by (17).

In the sequel we derive some analogous discrete entropy inequalities (with a possible remainder tending to 0) for the fully discrete, time semi-implicit (and fully implicit, i.e. backward Euler, as far as System (14) only is concerned) or segregated schemes (fully explicit regarding System (14) only) presented in Sect. 2, with a possible upwinding limited to that of the convection terms with respect to the material velocity. Note that the entropy inequalities that we obtain here apply to both the staggered schemes [14, 17, 21] and to the colocated scheme [25] which is also based on the internal energy; indeed the entropy depends only on the mass and internal energy which are scalar unknowns located at the center of the (primal) cells in both schemes, so System (14) involves only equations posed on the primal mesh.

Depending on the time and space discretization, we obtain three types of results:

- local entropy estimates, i.e. discrete analogues of (15), in which case the scheme is entropy stable,
- global entropy estimates, i.e. discrete analogues of:

$$\frac{d}{dt} \int_\Omega \eta(\rho, e) \, d\boldsymbol{x} \leq 0; \tag{21}$$

such a relation is a stability property of the scheme; this kind of relation was also proven in e.g. [7] for a higher order scheme for the 1D Euler equations;

- "weak local" entropy inequalities, i.e. results of the form:

$$\partial_t \eta(\rho, e) + \mathrm{div}\big[\eta(\rho, e) \, \boldsymbol{u}\big] + \widetilde{\mathcal{R}} \leq 0,$$

with $\widetilde{\mathcal{R}}$ tending to zero in a suitable sense with respect to the space and time discretization steps (or combination of both parameters), provided that the approximate solutions are controlled in reasonable norms, here, L^∞ and BV norms. Then a "Lax-consistency" property holds, of the form: a limit $(\bar{\rho}, \bar{\boldsymbol{u}}, \bar{e})$ of a convergent subsequence of approximate solutions given by the considered numerical scheme and bounded in the L^∞ and BV norms, satisfies the following weak entropy inequality:

$$-\int_0^T \int_\Omega \eta(\bar{\rho}, \bar{e}) \, \partial_t \varphi + \eta(\bar{\rho}, \bar{e}) \, \boldsymbol{u} \cdot \nabla \varphi \, d\boldsymbol{x} \, dt - \int_\Omega \eta(\bar{\rho}, \bar{e})(\boldsymbol{x}, 0) \, \varphi(\boldsymbol{x}, 0) \, d\boldsymbol{x} \leq 0,$$

$$\text{for any function } \varphi \in C_c^\infty\big([0, T) \times \bar{\Omega}\big), \varphi \geq 0. \tag{22}$$

In the sequel we address implicit schemes (Sect. 3.2) and segregated explicit schemes (Sect. 3.3). For implicit schemes, we first consider an upwind discretization for which we get a local discrete entropy inequality (Theorem 1), and then a MUSCL-like improvement of the discretization of the convection term in order to reduce the numerical diffusion, for which we only get a global entropy estimate and a weak local entropy inequality (Theorem 2). The case of explicit schemes is a little more tricky: we again consider the same two discretizations (i.e. upwind and MUSCL-like) but we first deal with the mass balance equation, then with the internal energy equation, and combine the results to address entropy inequalities.

3.1   Meshes and Discrete Norms

Let $\mathcal{M}$ be a mesh of the domain Ω, supposed to be regular in the usual sense of the finite element literature (see e.g. [6]). By $\mathcal{E}$ and $\mathcal{E}(K)$ we denote the set of all $(d-1)$-faces σ of the mesh and of the cell $K \in \mathcal{M}$ respectively, and we suppose that the number of the faces of a cell is bounded. The set of faces included in Ω (resp. in the boundary $\partial\Omega$) is denoted by $\mathcal{E}_{\text{int}}$ (resp. $\mathcal{E}_{\text{ext}}$); a face $\sigma \in \mathcal{E}_{\text{int}}$ separating the cells K and L is denoted by $\sigma = K|L$. For $K \in \mathcal{M}$ and $\sigma \in \mathcal{E}$, we denote by $|K|$ the measure of K and by $|\sigma|$ the $(d-1)$-measure of the face σ. The following quantities related to the mesh are used in the sequel:

$$h_{\mathcal{M}} = \max_{K \in \mathcal{M}} h_K \text{ with } h_K = \text{diam}(K), \qquad \underline{h}_{\mathcal{M}} = \min_{K \in \mathcal{M}} \frac{|K|}{\displaystyle\sum_{\sigma \in \mathcal{E}(K)} |\sigma|}. \tag{23}$$

$$C_{\mathcal{M}} = \max_{K \in \mathcal{M},\, (\sigma,\sigma') \in \mathcal{E}(K)^2} \frac{(|\sigma| + |\sigma'|)\, h_K}{|K|}, \qquad f_{\mathcal{M}} = \max_{K \in \mathcal{M}} \text{card } \mathcal{E}(K). \tag{24}$$

Let $(t_n)_{0 \le n \le N}$, with $0 = t_0 < t_1 < \dots < t_N = T$, define a partition of the time interval $(0, T)$, which we suppose uniform for the sake of simplicity, and let $\delta t = t_{n+1} - t_n$ for $0 \le n \le N-1$ be the (constant) time step.

The discrete pressure, density and the internal energy unknowns are associated with the cells of the mesh $\mathcal{M}$; they are denoted by:

$$\{p_K^n,\ \rho_K^n,\ e_K^n,\ K \in \mathcal{M},\ 0 \le n \le N\}.$$

In the estimates given below, we shall need some discrete norms that we now define.

Definition 1 (Discrete BV Semi-norms and Weak $L^1(0, T; (W_0^{1,+\infty})')$ Norm) For a family $(z_K^n)_{K \in \mathcal{M}, 0 \le n \le N} \subset \mathbb{R}$, let us define the following norms of the associated piecewise constant function z:

$$\|z\|_{\mathcal{T},t,\text{BV}} = \sum_{n=0}^{N} \sum_{K \in \mathcal{M}} |K|\, |z_K^{n+1} - z_K^n|,$$

$$\|z\|_{\mathcal{T},x,\text{BV}} = \sum_{n=0}^{N} \delta t \sum_{\sigma = K|L \in \mathcal{E}_{\text{int}}} |\sigma|\, |z_L^n - z_K^n|,$$

$$\|z\|_{-1,1,\star} = \sup_{\psi \in C_c^\infty([0,T] \times \Omega)} \frac{1}{\|\nabla \psi\|_\infty} \left[\sum_{n=0}^{N} \delta t \sum_{K \in \mathcal{M}} |K|\, z_K^n \psi_K^n \right],$$

$$\tag{25}$$

where ψ_K^n stands for $\psi(x_K, t_n)$, with x_K the mass center of K. Note that this latter weak norm is the discrete equivalent of the continuous dual norm of $v \in L^1(\Omega \times (0, T))$, defined by

$$\|v\|_{L^1(0,T;(W_0^{1,+\infty})')} = \sup_{\psi \in C_c^\infty([0, T) \times \Omega)} \frac{1}{\|\nabla \psi\|_\infty} \int_0^T \int_\Omega v\, \psi\, dx\, dt.$$

Some of the proofs below are based on the following convexity result [12, Lemma 2.3]. In its formulation, and throughout the paper, $[\![a, b]\!]$ stands for the interval $[\min(a, b), \max(a, b)]$, for any real numbers a and b.

Lemma 2 *Let φ be a strictly convex and continuously differentiable function over an open interval I of $\mathbb{R}$. Let $x_K \in I$ and $x_L \in I$ be two real numbers. Then the relation*

$$\varphi(x_K) + \varphi'(x_K)\,(x_{KL} - x_K) = \varphi(x_L) + \varphi'(x_L)\,(x_{KL} - x_L) \text{ if } x_K \neq x_L,$$

$$x_{KL} = x_K = x_L \text{ otherwise} \qquad (26)$$

uniquely defines the real number x_{KL} in $[\![x_K, x_L]\!]$.

Remark 1 (x_{KL} for $\varphi(z) = z^2$) Let us consider the specific function $\varphi(z) = z^2$. Then, an easy computation yields $x_{KL} = (x_K + x_L)/2$ i.e. the centered approximation.

3.2 Implicit Schemes

With the above notations, the space time discretization of System (14) reads:

For $K \in \mathcal{M}$, $0 \leq n \leq N - 1$,

$$\frac{|K|}{\delta t}(\rho_K^{n+1} - \rho_K^n) + \sum_{\sigma \in \mathcal{E}(K)} F_{K,\sigma}^{n+1} = 0, \qquad (27a)$$

$$\frac{|K|}{\delta t}(\rho_K^{n+1} e_K^{n+1} - \rho_K^n e_K^n) + \sum_{\sigma \in \mathcal{E}(K)} F_{K,\sigma}^{n+1} e_\sigma^{n+1} + p_K^{n+1} \sum_{\sigma \in \mathcal{E}(K)} |\sigma|\, u_{K,\sigma}^{n+1} \geq 0, \qquad (27b)$$

$$p_K^{n+1} = (\gamma - 1)\, \rho_K^{n+1}\, e_K^{n+1}, \qquad (27c)$$

where $F_{K,\sigma}^{n+1}$ is the mass flux through the face σ, e_σ^{n+1} is an approximation of the internal energy at the face σ, and $u_{K,\sigma}^{n+1}$ stands for an approximation of the normal velocity to the face σ; note that the velocity is solved in the full scheme by a space discretization of the momentum prediction and correction equations (9a)–(9b). Consistently with the boundary conditions, $u_{K,\sigma}^{n+1}$ vanishes on every external face. The mass flux $F_{K,\sigma}^{n+1}$ reads:

$$F_{K,\sigma}^{n+1} = |\sigma| \, \rho_\sigma^{n+1} u_{K,\sigma}^{n+1}, \tag{28}$$

where ρ_σ^{n+1} stands for an approximation of the density on σ. Throughout the paper, we suppose that ρ_K^n, e_K^n, ρ_σ^n and e_σ^n are positive, for any $K \in \mathcal{M}$, $\sigma \in \mathcal{E}_{\text{int}}$, $0 \leq n \leq N$, which is verified by the solutions of the schemes presented in [14, 16, 21, 22] (of course, with positive initial conditions for ρ and e).

The two following lemmas are straight forward consequences of Lemmas A1 and A2 in [21] and state discrete analogues of (18) and (19) respectively which are used to obtain the entropy inequalities.

Lemma 3 *Let $K \in \mathcal{M}$, n be such that $0 \leq n \leq N - 1$ and let us suppose that the discrete mass balance (27a) holds. Let φ be a twice continuously differentiable function defined over $(0, +\infty)$. Then*

$$\frac{|K|}{\delta t} \left[\varphi(\rho_K^{n+1}) - \varphi(\rho_K^n) \right] + \sum_{\sigma \in \mathcal{E}(K)} |\sigma| \, \varphi(\rho_\sigma^{n+1}) \, u_{K,\sigma}^{n+1}$$

$$+ \left[\rho_K^{n+1} \varphi'(\rho_K^{n+1}) - \varphi(\rho_K^{n+1}) \right] \sum_{\sigma \in \mathcal{E}(K)} |\sigma| \, u_{K,\sigma}^{n+1} + |K| \, (R_m)_K^{n+1} = 0, \tag{29}$$

with

$$|K| \, (R_m)_K^{n+1} = \frac{1}{2} \frac{|K|}{\delta t} \, \varphi''(\rho_K^{n+1/2}) \, (\rho_K^{n+1} - \rho_K^n)^2$$

$$+ \sum_{\sigma \in \mathcal{E}(K)} |\sigma| \left[\varphi(\rho_K^{n+1}) - \varphi(\rho_\sigma^{n+1}) + \varphi'(\rho_K^{n+1})(\rho_\sigma^{n+1} - \rho_K^{n+1}) \right] u_{K,\sigma}^{n+1},$$

where $\rho_K^{n+1/2} \in [\![\rho_K^n, \rho_K^{n+1}]\!]$.

Lemma 4 *Let $K \in \mathcal{M}$ and n be such that $0 \leq n \leq N - 1$. Let φ be a twice continuously differentiable function defined over $(0, +\infty)$. Then:*

$$\varphi'(e_K^{n+1}) \left[\frac{|K|}{\delta t}(\rho_K^{n+1} e_K^{n+1} - \rho_K^n e_K^n) + \sum_{\sigma \in \mathcal{E}(K)} F_{K,\sigma}^{n+1} e_\sigma^{n+1} \right] =$$

$$\frac{|K|}{\delta t} \left[\rho_K^{n+1} \varphi(e_K^{n+1}) - \rho_K^n \varphi(e_K^n) \right] + \sum_{\sigma \in \mathcal{E}(K)} F_{K,\sigma}^{n+1} \, \varphi(e_\sigma^{n+1}) + |K| \, (R_e)_K^{n+1}, \tag{30}$$

with

$$|K|\,(R_e)_K^{n+1} = \frac{1}{2}\frac{|K|}{\delta t}\rho_K^n\,\varphi''(e_K^{n+1/2})(e_K^{n+1} - e_K^n)^2$$

$$+ \sum_{\sigma\in\mathcal{E}(K)} F_{K,\sigma}^{n+1}\left[\varphi(e_K^{n+1}) - \varphi(e_\sigma^{n+1}) + \varphi'(e_K^{n+1})(e_\sigma^{n+1} - e_L^{n+1})\right],$$

where $e_K^{n+1/2} \in [\![e_K^n, e_K^{n+1}]\!]$.

Upwind Implicit Schemes In this section, we suppose that the convection fluxes are approximated with a first order upwind scheme, i.e., for $\sigma \in \mathcal{E}_{\text{int}}$, $\sigma = K|L$, $\rho_\sigma^{n+1} = \rho_K^{n+1}$ and $e_\sigma^{n+1} = e_K^{n+1}$ if $u_{K,\sigma} \geq 0$, $\rho_\sigma^{n+1} = \rho_L^{n+1}$ and $e_\sigma^{n+1} = e_L^{n+1}$ otherwise. In this case, the scheme (27) satisfies a local entropy estimate (i.e. a discrete analogue of Inequality (15)) which is stated in Theorem 1 below. Of course, this local entropy inequality also yields the global discrete inequality analogue to (21); furthermore, passing to the limit on the upwind implicit (or pressure correction) scheme applied the Euler equations, this local estimate also yields the Lax consistency, i.e. any limit $(\bar{\rho}, \bar{u}, \bar{e})$ of a convergent subsequence of approximate solutions satisfies the weak entropy inequality (22).

Theorem 1 (Discrete Entropy Inequality, Implicit Upwind Scheme) *Let η be defined by (17), and, for $0 \leq n \leq N - 1$, let $\eta_K^m = \eta(\rho_K^m, e_K^m)$ for $m = n$, $n + 1$ and $K \in \mathcal{M}$, and $\eta_\sigma^{n+1} = \eta(\rho_\sigma^{n+1}, e_\sigma^{n+1})$ for $\sigma \in \mathcal{E}_{\text{int}}$. Then any solution of the scheme (27) satisfies, for any $K \in \mathcal{M}$ and $0 \leq n \leq N - 1$:*

$$\frac{|K|}{\delta t}(\eta_K^{n+1} - \eta_K^n) + \sum_{\sigma\in\mathcal{E}(K)} |\sigma|\,\eta_\sigma^{n+1}u_{K,\sigma}^{n+1} \leq 0.$$

Proof Let φ be a twice continuously differentiable function. By Lemma 3, we get that (29) holds. For $\sigma \in \mathcal{E}_{\text{ext}}$, thanks to the boundary conditions, the convection fluxes vanish. For $0 \leq n \leq N-1$ and $K \in \mathcal{M}$, consider the term $(T_m)_{K,\sigma}^{n+1}$ associated to an internal face $\sigma = K|L$ in the remainder term $(R_m)_K^{n+1}$:

$$(T_m)_{K,\sigma}^{n+1} = \left[\varphi(\rho_K^{n+1}) - \varphi(\rho_\sigma^{n+1}) + \varphi'(\rho_K^{n+1})(\rho_\sigma^{n+1} - \rho_K^{n+1})\right]u_{K,\sigma}^{n+1}$$

$$= -\frac{1}{2}\varphi''(\rho_{\sigma,K}^{n+1})\,(\rho_\sigma^{n+1} - \rho_K^{n+1})^2 u_{K,\sigma}^{n+1},$$

where $\rho_{\sigma,K}^{n+1} \in [\![\rho_\sigma^{n+1}, \rho_K^{n+1}]\!]$. With the upwind choice, if $u_{K,\sigma}^{n+1} \geq 0$, $\rho_\sigma^{n+1} = \rho_K^{n+1}$ and $(T_m)_{K,\sigma}^{n+1}$ vanishes. If $u_{K,\sigma}^{n+1} < 0$ and φ'' is a non-negative function (*i.e.* φ is convex), $(T_m)_{K,\sigma}^{n+1}$ is non-negative and so is $(R_m)_K^{n+1}$, for any $K \in \mathcal{M}$. Since φ_ρ defined by (16) is indeed convex, Lemma 3 implies that any solution $\{\rho_K^n, K \in$

$\mathcal{M}, 0 \leq n \leq N\}$ to Eq. (27a) of the scheme satisfies, for $K \in \mathcal{M}$ and $0 \leq n \leq N - 1$:

$$\frac{|K|}{\delta t}\left[\varphi_\rho(\rho_K^{n+1}) - \varphi_\rho(\rho_K^n)\right] + \sum_{\sigma \in \mathcal{E}(K)} |\sigma|\,\varphi_\rho(\rho_\sigma^{n+1}) u_{K,\sigma}^{n+1}$$

$$+ \left[\rho_K^{n+1}\varphi_\rho'(\rho_K^{n+1}) - \varphi_\rho(\rho_K^{n+1})\right] \sum_{\sigma \in \mathcal{E}(K)} |\sigma|\, u_{K,\sigma}^{n+1} \leq 0. \qquad (31)$$

Now turning to Lemma 4, by similar arguments, the remainder term $(R_e)_K^{n+1}$ in (30) is nonnegative for any regular convex function φ, for any $K \in \mathcal{M}$ and $0 \leq n \leq N - 1$. Hence, since φ_e defined by Eq. (16) is convex, we get that any solution to (27b) satisfies:

$$\frac{|K|}{\delta t}\left[\rho_K^{n+1}\varphi_e(e_K^{n+1}) - \rho_K^n\varphi_e(e_K^n)\right]$$

$$+ \sum_{\sigma \in \mathcal{E}(K)} F_{K,\sigma}^{n+1}\,\varphi_e(e_\sigma^{n+1}) + \varphi_e'(e_K^{n+1})\,p_K^{n+1} \sum_{\sigma \in \mathcal{E}(K)} |\sigma|\, u_{K,\sigma}^{n+1} \leq 0. \qquad (32)$$

The desired relation is then obtained by summing the inequalities (31) and (32), using (20).

MUSCL-Like Schemes The aim of this section is to improve the approximation of the convection fluxes in (27a) and (27b) in order to reduce the numerical diffusion, while still satisfying an entropy inequality. This leads to a condition similar to the limitation procedure which is the core of a MUSCL procedure [41]; indeed, in order to yield an entropy inequality (instead of, for a MUSCL technique, to yield a maximum principle), the approximation of the unknowns at the face must be "sufficiently close to" the upwind approximation. The entropy inequality is then obtained only in the weak sense. The technique to reach this result consists in splitting the remainder terms appearing in Lemmas 3 and 4 in two parts: the first one is non-negative under some condition for the face approximation (hence the above mentioned limitation requirement); the second one is conservative and can be bounded in a discrete negative Sobolev norm (this explains why the entropy estimate is only a weak one).

Let φ_ρ and φ_e be the functions defined by (16) and let $\sigma \in \mathcal{E}_{\mathrm{int}}$, $\sigma = K|L$; by Lemma 2, there exists a unique $\rho_{KL}^{n+1} \in [\![\rho_K^{n+1}, \rho_L^{n+1}]\!]$ and $e_{KL}^{n+1} \in [\![e_K^{n+1}, e_L^{n+1}]\!]$ such that

$$\varphi_\rho(\rho_K^{n+1}) + \varphi_\rho'(\rho_K^{n+1})\left[\rho_{KL}^{n+1} - \rho_K^{n+1}\right] = \varphi_\rho(\rho_L^{n+1}) + \varphi_\rho'(\rho_L^{n+1})\left[\rho_{KL}^{n+1} - \rho_L^{n+1}\right],$$

$$(33a)$$

$$\varphi_e(e_K^{n+1}) + \varphi_e'(e_K^{n+1}) \left[e_{KL}^{n+1} - e_K^{n+1} \right] = \varphi_e(e_L^{n+1}) + \varphi_e'(e_L^{n+1}) \left[e_{KL}^{n+1} - e_L^{n+1} \right].$$
(33b)

Entropy estimates are obtained in Theorem 2 under the following conditions:

$$\rho_\sigma^{n+1} \in [\![\rho_K^{n+1}, \ \rho_{KL}^{n+1}]\!] \text{ if } u_{K,\sigma}^{n+1} \geq 0, \quad \rho_\sigma^{n+1} \in [\![\rho_L^{n+1}, \ \rho_{KL}^{n+1}]\!] \text{ otherwise}, \quad (34a)$$

$$e_\sigma^{n+1} \in [\![e_K^{n+1}, \ e_{KL}^{n+1}]\!] \text{ if } u_{K,\sigma}^{n+1} \geq 0, \quad e_\sigma^{n+1} \in [\![e_L^{n+1}, \ e_{KL}^{n+1}]\!] \text{ otherwise}, \quad (34b)$$

where ρ_{KL}^{n+1} and e_{KL}^{n+1} are defined by (33). Note that these conditions are satisfied by the upwind scheme (27). They may be seen as an additional constraint to be added to the limitation of a MUSCL-like procedure (see also the conclusion of the last section of this paper).

Theorem 2 (Entropy Inequalities, Implicit MUSCL-Like Scheme) *Let us assume that, for $\sigma \in \mathcal{E}_{int}$, $\sigma = K|L$ and for $0 \leq n \leq N - 1$, the approximate density ρ_σ^{n+1} and internal energy e_σ^{n+1} in the numerical mass fluxes (28) and in the internal energy balance (27b) satisfy the conditions (34).*

Then any solution of the scheme (27) satisfies, for any $K \in \mathcal{M}$ and $0 \leq n \leq N - 1$:

$$\frac{|K|}{\delta t} (\eta_K^{n+1} - \eta_K^n) + \sum_{\sigma \in \mathcal{E}(K)} |\sigma| \, \eta_\sigma^{n+1} u_{K,\sigma}^{n+1} + |K| \, (\delta R_\eta)_K^{n+1} \leq 0,$$

where the remainder term δR_η satisfies $\sum_{K \in \mathcal{M}} |K| \, (\delta R_\eta)_K^{n+1} = 0$ so that, integrating in space (i.e. summing over the cells), the following global discrete entropy estimate holds for $0 \leq n \leq N - 1$:

$$\sum_{K \in \mathcal{M}} |K| \, \eta_K^{n+1} \leq \sum_{K \in \mathcal{M}} |K| \, \eta_K^n.$$

In addition, let us suppose that there exists $M > 0$ such that $\rho_K^n \leq M$, $1/\rho_K^n \leq M$, $e_K^n \leq M$, $1/e_K^n \leq M$ and $|u_{K,\sigma}^n| \leq M$ for $K \in \mathcal{M}$, $\sigma \in \mathcal{E}(K)$ and $0 \leq n \leq N$, and let us define the quantities $|\varphi_\rho'|_\infty = \max(|\varphi_\rho'(1/M)|, \ |\varphi_\rho'(M)|)$ and $|\varphi_e'|_\infty = \max(|\varphi_e'(1/M)|, \ |\varphi_e'(M)|)$. Then the remainder term δR_m satisfies the following bound:

$$\|\delta R_m\|_{-1,1,\star} \leq 3 \, M \left(|\varphi_\rho'|_\infty \, \|\rho\|_{\mathcal{T},x,BV} + M \, |\varphi_e'|_\infty \, \|e\|_{\mathcal{T},x,BV} \right) h_\mathcal{M}. \quad (35)$$

Therefore, a Lax-consistency property holds; more precisely, any limit $(\bar{\rho}, \bar{u}, \bar{e})$ of a converging sequence of approximate solutions bounded in the L^∞ and BV norms satisfies (22).

Proof Let $(\delta\varphi_\rho)_\sigma^{n+1}$ be defined by:

$$(\delta\varphi_\rho)_\sigma^{n+1} = \varphi_\rho(\rho_K^{n+1}) - \varphi_\rho(\rho_\sigma^{n+1}) + \varphi_\rho'(\rho_K^{n+1})\left[\rho_{KL}^{n+1} - \rho_K^{n+1}\right]$$

$$+ \frac{1}{2}\left[\varphi_\rho'(\rho_K^{n+1}) + \varphi_\rho'(\rho_L^{n+1})\right]\left[\rho_\sigma^{n+1} - \rho_{KL}^{n+1}\right]. \quad (36)$$

By Lemma 3, (29) holds; an easy computation shows that the term associated to the face σ in the expression of the remainder term $(R_m)_K^{n+1}$ satisfies:

$$(F_m)_{K,\sigma}^{n+1} = |\sigma|\left[\varphi_\rho(\rho_K^{n+1}) - \varphi_\rho(\rho_\sigma^{n+1}) + \varphi_\rho'(\rho_K^{n+1})(\rho_\sigma^{n+1} - \rho_K^{n+1})\right]u_{K,\sigma}^{n+1}$$

$$= |\sigma|\,(\delta\varphi_\rho)_\sigma^{n+1}\,u_{K,\sigma}^{n+1} + (F_m^R)_{K,\sigma}^{n+1}$$

with $(F_m^R)_{K,\sigma}^{n+1} = |\sigma|\frac{1}{2}\left[\varphi_\rho'(\rho_K^{n+1}) - \varphi_\rho'(\rho_L^{n+1})\right](\rho_\sigma^{n+1} - \rho_{KL}^{n+1})\,u_{K,\sigma}^{n+1}$. Thanks to the assumption (34a), since φ_ρ' is an increasing function, $(F_m^R)_{K,\sigma}^{n+1} \geq 0$. Let us define $(\delta R_m)_K^{n+1}$, $K \in \mathcal{M}, 0 \leq n \leq N-1$ by:

$$|K|\,(\delta R_m)_K^{n+1} = \sum_{\sigma \in \mathcal{E}(K)} |\sigma|\,(\delta\varphi_\rho)_\sigma^{n+1}\,u_{K,\sigma}^{n+1}. \quad (37)$$

Then, under assumption (34a), we get:

$$\frac{|K|}{\delta t}\left[\varphi_\rho(\rho_K^{n+1}) - \varphi_\rho(\rho_K^n)\right] + \sum_{\sigma \in \mathcal{E}(K)} |\sigma|\,\varphi_\rho(\rho_\sigma^{n+1})u_{K,\sigma}^{n+1}$$

$$+ \left[\rho_K^{n+1}\varphi_\rho'(\rho_K^{n+1}) - \varphi_\rho(\rho_K^{n+1})\right]\sum_{\sigma \in \mathcal{E}(K)} |\sigma|\,u_{K,\sigma}^{n+1} + |K|\,(\delta R_m)_K^{n+1} \leq 0. \quad (38)$$

Let us prove that δR_m satisfies:

$$\|\delta R_m\|_{-1,1,\star} \leq 3\,M\,|\varphi_\rho'|_\infty\,\|\rho\|_{\mathcal{T},x,\mathrm{BV}}\,h_\mathcal{M}. \quad (39)$$

Indeed, since both ρ_σ^{n+1} and ρ_{KL}^{n+1} lie in the interval $[\![\rho_K^{n+1},\ \rho_L^{n+1}]\!]$, we have by convexity of φ_ρ:

$$|(\delta\varphi_\rho)_\sigma^{n+1}| \leq 3\max\left(|\varphi_\rho'(\rho_K^{n+1})|, |\varphi_\rho'(\rho_L^{n+1})|\right)|\rho_K^{n+1} - \rho_L^{n+1}|.$$

Let ψ be a function of $C_c^\infty(\Omega \times (0, T))$. We have, thanks to the conservativity of the remainder term:

$$T = \sum_{n=0}^{N-1} \delta t \sum_{K \in \mathcal{M}} |K| \, (\delta R_m)_K^{n+1} \, \psi_K^{n+1}$$

$$= \sum_{n=0}^{N-1} \delta t \sum_{\sigma = K|L \in \mathcal{E}_{\text{int}}} |\sigma| \, (\delta \varphi_\rho)_\sigma^{n+1} \, (\psi_K^{n+1} - \psi_L^{n+1}) \, u_{K,\sigma}.$$

Therefore,

$$|T| \leq 3 \, |\varphi_\rho'|_\infty \, M \, \left[\|\nabla \psi\|_\infty \right] h_{\mathcal{M}} \sum_{n=0}^{N-1} \delta t \sum_{\sigma = K|L \in \mathcal{E}_{\text{int}}} |\sigma| \, |\rho_K^{n+1} - \rho_L^{n+1}|,$$

which concludes the proof of (39).

Following the same line of thought for the internal energy balance, let $(\delta \varphi_e)_\sigma^{n+1}$ be defined by:

$$(\delta \varphi_e)_\sigma^{n+1} = \varphi_e(e_K^{n+1}) - \varphi_e(e_\sigma^{n+1}) + \varphi_e'(e_K^{n+1}) \left[e_{KL}^{n+1} - e_K^{n+1} \right]$$

$$+ \frac{1}{2} \left[\varphi_e'(e_K^{n+1}) + \varphi_e'(e_L^{n+1}) \right] \left[e_\sigma^{n+1} - e_{KL}^{n+1} \right], \qquad (40)$$

and $(\delta R_e)_K^{n+1}$ the remainder term given by:

$$|K| \, (\delta R_e)_K^{n+1} = \sum_{\sigma \in \mathcal{E}(K)} (\delta \varphi_e)_\sigma^{n+1} \, F_{K,\sigma}^{n+1}. \qquad (41)$$

Thanks to the assumption (34b) we get:

$$\frac{|K|}{\delta t} \left[\rho_K^{n+1} \varphi_e(e_K^{n+1}) - \rho_K^n \varphi_e(e_K^n) \right] + \sum_{\sigma \in \mathcal{E}(K)} \varphi_e(e_\sigma^{n+1}) F_{K,\sigma}^{n+1}$$

$$+ \varphi_e'(e_K^{n+1}) p_K^{n+1} \sum_{\sigma \in \mathcal{E}(K)} |\sigma| \, u_{K,\sigma}^{n+1} + |K| \, (\delta R_e)_K^{n+1} \leq 0. \qquad (42)$$

In addition, δR_e satisfies the following inequality:

$$\|\delta R_e\|_{-1,1,\star} \leq 3 \, M^2 \, |\varphi_e'|_\infty \, \|e\|_{\mathcal{T},x,\text{BV}} \, h_{\mathcal{M}}. \qquad (43)$$

Combining the inequalities (38) and (42) and thanks to (20), (39) and (43) concludes the proof of the theorem.

3.3 Explicit Schemes

The general form of the discrete analogue of System (14) for an explicit scheme reads:

For $K \in \mathcal{M}$, $0 \le n \le N - 1$,

$$\frac{|K|}{\delta t}(\rho_K^{n+1} - \rho_K^n) + \sum_{\sigma \in \mathcal{E}(K)} F_{K,\sigma}^n = 0, \tag{44a}$$

$$\frac{|K|}{\delta t}(\rho_K^{n+1} e_K^{n+1} - \rho_K^n e_K^n) + \sum_{\sigma \in \mathcal{E}(K)} F_{K,\sigma}^n e_\sigma^n + p_K^n \sum_{\sigma \in \mathcal{E}(K)} |\sigma| u_{K,\sigma}^n \ge 0, \tag{44b}$$

$$p_K^n = (\gamma - 1) \rho_K^n e_K^n, \tag{44c}$$

where the numerical mass flux $F_{K,\sigma}^n$ is still defined by (28).

Let ρ_{KL}^n (resp. e_{KL}^n) be the real number defined by Eq. (26) with $x_K = \rho_K^n$ (resp. $x_K = e_K^n$) and $x_L = \rho_L^n$ (resp. $x_L = e_L^n$) and $\varphi = \varphi_\rho$ (resp. $\varphi = \varphi_e$) and let us assume that for $\sigma \in \mathcal{E}_{\text{int}}$, $\sigma = K|L$ and for $0 \le n \le N - 1$,

$$\rho_\sigma^n \in [\![\rho_K^n, \, \rho_{KL}^n]\!] \text{ if } u_{K,\sigma}^n \ge 0, \qquad \rho_\sigma^n \in [\![\rho_L^n, \, \rho_{KL}^n]\!] \text{ otherwise.} \tag{45}$$

$$e_\sigma^n \in [\![e_K^n, \, e_{KL}^n]\!] \text{ if } u_{K,\sigma}^n \ge 0, \qquad e_\sigma^n \in [\![e_L^n, \, e_{KL}^n]\!] \text{ otherwise.} \tag{46}$$

With these two conditions, Theorem 3 below yields a weak discrete entropy inequality, in the sense that a remainder term exists which tends to 0 (under some conditions) with the mesh and time steps, but its sign is unknown. However, in the case of an upwind approximation of the density and the internal energy on the faces of the mesh (note that (45) and (46) are satisfied for such approximations), a local discrete entropy inequality can be obtained under the following additional conditions.

1. First, the normal face velocities $u_{K,\sigma}$ in the mass flux (28) are assumed to be either

   – computed from a discrete velocity field $\boldsymbol{u}$:

   $$u_{K,\sigma} = \boldsymbol{u}_\sigma \cdot \boldsymbol{n}_{K,\sigma}, \tag{47}$$

   where $\boldsymbol{n}_{K,\sigma}$ is the unit normal vector to σ outward K and $\boldsymbol{u}_\sigma$ is an approximation of the velocity at the face, which may be the discrete unknown itself (when the velocity degrees of freedom are those of a non conforming Crouzeix-Raviart or Rannacher-Turek approximation, see e.g. [14]) or an interpolation (for instance, for a colocated arrangement of the unknowns, as in [25]).

- the unknown themselves in the case of the staggered MAC scheme, since only the normal velocity is approximated in this case, see e.g. [14].

For $1 \leq r$, we then define the following discrete norm:

$$\|u\|^r_{L^r(0,T;W^{1,r}_{\mathcal{M}})} = \sum_{i=1}^{d} \sum_{n=0}^{N} \delta t \sum_{K \in \mathcal{M}} \sum_{(\sigma,\sigma') \in \mathcal{E}^{(i)}(K)^2} |K| \left(\frac{u^n_{\sigma,i} - u^n_{\sigma',i}}{h_K} \right)^r, \quad (48)$$

where $\mathcal{E}^{(i)}(K) = \mathcal{E}(K)$ for the Crouzeix-Raviart or Rannacher-Turek case and $\mathcal{E}^{(i)}(K)$ is restricted to the two faces of K perpendicular to the ith vector of the canonical basis of $\mathbb{R}^d$ in the case of the MAC scheme.

Remark 2 (Discrete $L^r(W^{1,r})$ Norm of the Velocity) It is reasonable to suppose that, under regularity assumptions of the mesh whose precise statement depends the space approximation at hand, this norm is equivalent to the standard finite-volume discrete $L^r(0,T;W^{1,r})$ norm [10]; it is indeed true for usual cells (in particular, with a bounded number of faces) for staggered discretizations and for a convex interpolation of the velocity at the faces for colocated schemes.

2. Second, the following CFL conditions hold:

$$\delta t \leq \frac{|K|}{\displaystyle\sum_{\sigma \in \mathcal{E}(K)} \frac{\varphi''_\rho(\tilde{\rho}^{n+1/2}_K)^2}{\varphi''_\rho(\rho^n_{K,\sigma})} |\sigma| (u^n_{K,\sigma})^-}, \quad (49)$$

$$\delta t \leq \frac{\varphi''_e(e^{n+1/2}_K) |K| \rho^{n+1}_K}{\displaystyle\sum_{\sigma \in \mathcal{E}(K)} \frac{\varphi''_e(\tilde{e}^{n+1/2}_K)^2}{\varphi''_e(e^n_{K,\sigma})} (F^n_{K,\sigma})^-}. \quad (50)$$

where $\tilde{\rho}^{n+1/2}_K \in [\![\rho^n_K, \rho^{n+1}_K]\!]$, $\rho^n_{K,\sigma} \in [\![\rho^n_K, \rho^n_L]\!]$, $\tilde{e}^{n+1/2}_K \in [\![e^n_K, e^{n+1}_K]\!]$ and $e^n_{K,\sigma} \in [\![e^n_K, e^n_L]\!]$ are defined by:

$$\varphi''_\rho(\tilde{\rho}^{n+1/2}_K) (\rho^{n+1}_K - \rho^n_K)^2 = \varphi'_\rho(\rho^{n+1}_K) - \varphi'_\rho(\rho^n_K), \quad (51)$$

$$\varphi''_\rho(\rho^n_{K,\sigma}) (\rho^n_K - \rho^n_L)^2 = \varphi_\rho(\rho^n_L) - \varphi_\rho(\rho^n_K) - \varphi'_\rho(\rho^n_K)(\rho^n_K - \rho^n_L), \quad (52)$$

$$\varphi''_e(\tilde{e}^{n+1/2}_K) (e^{n+1}_K - e^n_K)^2 = \varphi'_e(e^{n+1}_K) - \varphi'_e(e^n_K), \quad (53)$$

$$\varphi''_e(e^n_{K,\sigma}) (e^n_K - e^n_L)^2 = \varphi_e(e^n_L) - \varphi_e(e^n_K) - \varphi'_e(e^n_K)(e^n_K - e^n_L). \quad (54)$$

Theorem 3 (Discrete Entropy Inequalities, Explicit Schemes) *Let ρ and e satisfy the relations of the scheme* (44). *Let $M \geq 1$ and let us suppose that $\rho_K^n \leq M$, $1/\rho_K^n \leq M$, $e_K^n \leq M$, $1/e_K^n \leq M$ and $|u_{K,\sigma}| \leq M$, for $K \in \mathcal{M}$, $\sigma \in \mathcal{E}(K)$ and $0 \leq n \leq N$. Assume that the discretization of the convection term in (44a) and (44b) satisfies the assumptions (45) and (46) respectively. Let η be defined by (17). Then any solution of the scheme (44) satisfies, for any $K \in \mathcal{M}$ and $0 \leq n \leq N - 1$:*

$$\frac{|K|}{\delta t}(\eta_K^{n+1} - \eta_K^n) + \sum_{\sigma \in \mathcal{E}(K)} |\sigma|\, \eta_\sigma^n u_{K,\sigma}^n + |K|\, (R_\eta)_K^n \leq 0,$$

where $R_\eta = R_{\eta,1} + R_{\eta,2}$ with:

$$\|R_{\eta,1}\|_{-1,1,\star} \leq 3M \left(|\varphi_\rho'|_\infty \|\rho\|_{\mathcal{T},x,\mathrm{BV}} + M\, |\varphi_e'|_\infty \|e\|_{\mathcal{T},x,\mathrm{BV}} \right) h_\mathcal{M},$$

$$\|R_{\eta,2}\|_{L^1} \leq M^2 \left(|\varphi_\rho''|_\infty \|\rho\|_{\mathcal{T},t,\mathrm{BV}} + |\varphi_e''|_\infty \|e\|_{\mathcal{T},t,\mathrm{BV}} \right) \frac{\delta t}{\underline{h}_\mathcal{M}},$$

where $\underline{h}_\mathcal{M}$ is defined by (23), $|\varphi_\rho'|_\infty = \max(|\varphi_\rho'(1/M)|,\ |\varphi_\rho'(M)|)$, $|\varphi_e'|_\infty = \max(|\varphi_e'(1/M)|,\ |\varphi_e'(M)|)$, *and $|\varphi_\rho''|_\infty$ and $|\varphi_e''|_\infty$ denote the maximum value taken by φ_ρ'' and φ_e'' respectively on the interval $[1/M,\, M]$.*

Moreover, if the discretization of the convection term in (44a) *and* (44b) *is upwind, and if the normal face velocities $u_{K,\sigma}$ satisfy* (47), *under the CFL conditions* (49) *and* (50), *we also have (with a different expression for R_η):*

$$\|R_\eta\|_{L^1} \leq f_\mathcal{M}\, C_\mathcal{M}\, M^{(2q-1)/q}\, |\varphi_\rho''|_\infty \|\rho\|_{\mathcal{T},t,\mathrm{BV}}^{\frac{1}{q}} \|u\|_{L^{q'}(0,T;W_\mathcal{M}^{1,q'})} \delta t^{\frac{1}{q}}. \tag{55}$$

where $q \geq 1$, $q' \geq 1$ and $\dfrac{1}{q} + \dfrac{1}{q'} = 1$, $f_\mathcal{M}$ and $C_\mathcal{M}$ are defined by (24).

Proof The results are obtained by applying the propositions 1 and 2 below with $\varphi = \varphi_\rho$ and $\varphi = \varphi_e$ respectively.

The aim of the following proposition is to derive a discrete analogue of Relation (18).

Proposition 1 (Discrete Renormalized Forms of the Mass Balance Equation) *Let φ be a twice continuously differentiable convex function from $(0, +\infty)$ to $\mathbb{R}$, and let ρ satisfy* (44a). *Let $M \geq 1$ and let us suppose that $\rho_K^n \leq M$, $1/\rho_K^n \leq M$ and $|u_{K,\sigma}| \leq M$, for $K \in \mathcal{M}$, $\sigma \in \mathcal{E}(K)$ and $0 \leq n \leq N$. Let $|\varphi'|_\infty = \max(|\varphi'(1/M)|,\ |\varphi'(M)|)$ and $|\varphi''|_\infty$ be the maximum value taken by φ'' on the* *interval* $[1/M,\ M]$. *Assume that* ρ_σ^n *satisfies* (45). *Then the following inequality holds:*

$$\frac{|K|}{\delta t}\big[\varphi(\rho_K^{n+1}) - \varphi(\rho_K^n)\big] + \sum_{\sigma \in \mathcal{E}(K)} |\sigma|\varphi(\rho_\sigma^n)u_{K,\sigma}^n$$

$$+ \big(\varphi'(\rho_K^n)\rho_K^n - \varphi(\rho_K^n)\big)\Big[\sum_{\sigma \in \mathcal{E}(K)} |\sigma|\,u_{K,\sigma}^n\Big] + |K|\,(R_\rho)_K^{n+1} \le 0, \qquad (56)$$

where the remainder $(R_\rho)_K^{n+1} = (R_{\rho,1})_K^{n+1} + (R_{\rho,2})_K^{n+1}$ *with:*

$$\|R_{\rho,1}\|_{-1,1,\star} \le 3M\,|\varphi'|_\infty\,\|\rho\|_{\mathcal{T},x,BV}\,h_\mathcal{M},$$

$$\|R_{\rho,2}\|_{L^1} \le M^2\,|\varphi''|_\infty\,\|\rho\|_{\mathcal{T},t,BV}\,\frac{\delta t}{\underline{h}_\mathcal{M}},$$

where $\underline{h}_\mathcal{M}$ *is defined by* (23).

Assume furthermore that the normal face velocities $u_{K,\sigma}$ *satisfy* (47), *that the discretization of the convection term in* (44a) *is upwind and that the CFL condition* (49) *holds with* φ *instead of* φ_ρ. *Then* (56) *still holds (with a different expression for* R_ρ) *and:*

$$\|R_\rho\|_{L^1} \le f_\mathcal{M}\,C_\mathcal{M}\,M^{(2q-1)/q}\,|\varphi''|_\infty\,\|\rho\|_{\mathcal{T},t,BV}^{1/q}\,\|\boldsymbol{u}\|_{L^{q'}(0,T;W_\mathcal{M}^{1,q'})}\,\delta t^{1/q},$$

where $q \ge 1,\ q' \ge 1,\ \dfrac{1}{q} + \dfrac{1}{q'} = 1$, $\|\cdot\|_{L^{q'}(0,T;W_\mathcal{M}^{1,q'})}$ *is defined by* (48), $f_\mathcal{M}$ *and* $C_\mathcal{M}$ *are defined by* (24) *and* C *only depends on the maximal number of faces of the mesh cells.*

Proof Mimicking the formal computation performed at the continuous level, let us multiply (44a) by $\varphi'(\rho_K^{n+1})$. We get:

$$\varphi'(\rho_K^{n+1})\Big[\frac{|K|}{\delta t}(\rho_K^{n+1} - \rho_K^n) + \sum_{\sigma \in \mathcal{E}(K)} F_{K,\sigma}^n\Big] = (T_1)_K^{n+1} + (T_2)_K^{n+1} + |K|\,R_K^{n+1} = 0,$$

with

$$(T_1)_K^{n+1} = \varphi'(\rho_K^{n+1})\frac{|K|}{\delta t}(\rho_K^{n+1} - \rho_K^n), \quad (T_2)_K^{n+1} = \varphi'(\rho_K^n)\sum_{\sigma \in \mathcal{E}(K)} F_{K,\sigma}^n,$$

$$|K|\,R_K^{n+1} = \big(\varphi'(\rho_K^{n+1}) - \varphi'(\rho_K^n)\big)\sum_{\sigma \in \mathcal{E}(K)} F_{K,\sigma}^n. \qquad (57)$$

By a Taylor expansion, there exists $\rho_K^{n+1/2} \in [\![\rho_K^n, \rho_K^{n+1}]\!]$ such that:

$$(T_1)_K^{n+1} = \frac{|K|}{\delta t} \left[\varphi(\rho_K^{n+1}) - \varphi(\rho_K^n) \right] + |K| (R_1)_K^{n+1},$$

$$\text{with } (R_1)_K^{n+1} = \frac{1}{2\delta t} \varphi''(\rho_K^{n+1/2}) (\rho_K^{n+1} - \rho_K^n)^2 \geq 0. \tag{58}$$

The term $(T_2)_K^{n+1}$ reads:

$$(T_2)_K^{n+1} = \sum_{\sigma \in \mathcal{E}(K)} |\sigma| \varphi(\rho_\sigma^n) u_{K,\sigma}^n$$

$$+ \left(\varphi'(\rho_K^n)\rho_K^n - \varphi(\rho_K^n) \right) \sum_{\sigma \in \mathcal{E}(K)} |\sigma| u_{K,\sigma}^n + |K| (R_2)_K^{n+1}, \tag{59}$$

with

$$|K| (R_2)_K^{n+1} = \sum_{\sigma \in \mathcal{E}(K)} |\sigma| \left[\varphi(\rho_K^n) + \varphi'(\rho_K^n)(\rho_\sigma^n - \rho_K^n) - \varphi(\rho_\sigma^n) \right] u_{K,\sigma}^n.$$

Thanks to assumption (45), the remainder R_2 is a sum of a non-negative part and a term tending to zero; indeed there exists δR_2 such that:

$$R_2 \geq \delta R_2 \text{ and } \|\delta R_2\|_{-1,1,\star} \leq 3M \, |\varphi'|_\infty \, \|\rho\|_{\mathcal{T},x,BV} \, h_{\mathcal{M}}.$$

This result is obtained by adapting the proof of the implicit case (indeed, up to a change of time exponents at the right-hand side from n to $n + 1$, the expression of $(R_2)_K^{n+1}$ is the same than the second term of $(R_m)_K^{n+1}$ in the expression (29), and the computation from Relation (36) up to the end of the proof of (39) may be reproduced, still with the same change of time exponents).

Let us now prove that the remainder term $R = (R_K^n)_{K \in \mathcal{M}}^{n=0,\ldots,M}$ defined by (57) satisfies:

$$\|R\|_{L^1} = \sum_{n=0}^{N-1} \delta t \sum_{K \in \mathcal{M}} |K| R_K^{n+1} \leq M^2 \, |\varphi''|_\infty \, \|\rho\|_{\mathcal{T},t,BV} \, \frac{\delta t}{h_{\mathcal{M}}}. \tag{60}$$

Indeed, for $K \in \mathcal{M}$ and $0 \leq n \leq N$, we get:

$$|K| R_K^{n+1} = \left(\varphi'(\rho_K^{n+1}) - \varphi'(\rho_K^n) \right) \sum_{\sigma \in \mathcal{E}(K)} F_{K,\sigma}^n \tag{61}$$

$$= \varphi''(\tilde{\rho}_K^{n+1/2})(\rho_K^{n+1} - \rho_K^n) \sum_{\sigma \in \mathcal{E}(K)} |\sigma| \, \rho_\sigma^n u_{K,\sigma}^n, \tag{62}$$

where $\tilde{\rho}_K^{n+1/2}$ is defined by (51). Thus,

$$\|R\|_{L^1} = \sum_{n=0}^{N-1} \delta t \sum_{K \in \mathcal{M}} |K|\, R_K^{n+1} \leq |\varphi''|_\infty\, M^2 \sum_{n=0}^{N-1} \delta t \Big(\sum_{K \in \mathcal{M}} |\sigma| \Big) |\rho_K^{n+1} - \rho_K^n|,$$

which yields (60).

Let us now turn to the case where the discrete normal velocities satisfy (47) and the discretization of the density at the face ρ_σ^n is upwind; in this case the remainder R_2 defined by (3.3) satisfies:

$$|K|\, (R_2)_K^{n+1} = \sum_{\sigma=K|L} \frac{1}{2} |\sigma|\, \varphi''(\rho_{K,\sigma}^n)\, (\rho_K^n - \rho_L^n)^2 (u_{K,\sigma}^n)^-, \tag{63}$$

where $\rho_{K,\sigma}^n$ is defined by (52). Therefore, R_2 is non-negative. Starting from Eq. (61), we may now reformulate the remainder term R_K^{n+1} as $R_K^{n+1} = (R_{01})_K^{n+1} + (R_{02})_K^{n+1}$ with:

$$\begin{aligned}
|K|\, (R_{01})_K^{n+1} &= \varphi''(\tilde{\rho}_K^{n+1/2})(\rho_K^{n+1} - \rho_K^n)\, \rho_K^n \Big[\sum_{\sigma \in \mathcal{E}(K)} |\sigma|\, u_{K,\sigma}^n \Big], \\
|K|\, (R_{02})_K^{n+1} &= \varphi''(\tilde{\rho}_K^{n+1/2})(\rho_K^{n+1} - \rho_K^n) \Big[\sum_{\sigma \in \mathcal{E}(K)} |\sigma|(\rho_\sigma^n - \rho_K^n) u_{K,\sigma}^n \Big].
\end{aligned} \tag{64}$$

By Young's inequality, the second term may be estimated as follows:

$$\begin{aligned}
|K|\,|(R_{02})_K^{n+1}| \leq\ & \frac{1}{2} \sum_{\sigma \in \mathcal{E}(K)} |\sigma|\, \varphi''(\rho_{K,\sigma}^n)\, (u_{K,\sigma}^n)^-\, (\rho_K^n - \rho_L^n)^2 \\
& + \frac{1}{2} \sum_{\sigma \in \mathcal{E}(K)} |\sigma|\, \frac{\varphi''(\tilde{\rho}_K^{n+1/2})^2}{\varphi''(\rho_{K,\sigma}^n)}\, (u_{K,\sigma}^n)^-\, (\rho_K^{n+1} - \rho_K^n)^2.
\end{aligned}$$

Therefore, in view of the expressions (58) and (63) of $(R_1)_K^n$ and $(R_2)_K^n$ respectively, we get $(R_1)_K^{n+1} + (R_2)_K^{n+1} + (R_{02})_K^{n+1} \geq 0$ under the CFL condition (49) (with φ instead of φ_ρ). Let us now show that

$$\|R_{01}\|_{L^1} \leq f_{\mathcal{M}}\, C_{\mathcal{M}}\, M^{(2q-1)/q}\, |\varphi''|_\infty\, \|\rho\|_{\mathcal{T},t,\mathrm{BV}}^{1/q}\, \|u\|_{L^{q'}(0,T;W_{\mathcal{M}}^{1,q'})}\, \delta t^{1/q},$$
$$\tag{65}$$

with $q \geq 1$, $q' \geq 1$ and $\dfrac{1}{q} + \dfrac{1}{q'} = 1$, where $f_{\mathcal{M}}$ and $C_{\mathcal{M}}$ is defined by (24). To this purpose, we first observe that, in the Crouzeix-Raviart or Ranncher-Turek case, since $\sum_{\sigma \in \mathcal{E}(K)} |\sigma| \, \boldsymbol{n}_{K,\sigma} = 0$, we may write:

$$\sum_{\sigma \in \mathcal{E}(K)} |\sigma| \, u_{K,\sigma}^n = \sum_{\sigma \in \mathcal{E}(K)} |\sigma| \, \boldsymbol{u}_\sigma^n \cdot \boldsymbol{n}_{K,\sigma} = \sum_{\sigma \in \mathcal{E}(K)} |\sigma| \, (\boldsymbol{u}_\sigma^n - \boldsymbol{u}_K^n) \cdot \boldsymbol{n}_{K,\sigma},$$

where $\boldsymbol{u}_K^n$ stands for the mean value of the normal face velocities $(\boldsymbol{u}_\sigma^n)_{\sigma \in \mathcal{E}(K)}$. In the MAC case, we have $\sum_{\sigma \in \mathcal{E}^{(i)}(K)} |\sigma| \, u_{K,\sigma}^n = 0$ for $1 \le i \le d$, and thus

$$\sum_{\sigma \in \mathcal{E}(K)} |\sigma| \, u_{K,\sigma}^n = \sum_{i=1}^d \sum_{\sigma \in \mathcal{E}^{(i)}(K)} |\sigma| \, u_{\sigma,i}^n \, \boldsymbol{j}^{(i)} \cdot \boldsymbol{n}_{K,\sigma} =$$

$$\sum_{i=1}^d \sum_{\sigma \in \mathcal{E}^{(i)}(K)} |\sigma| \, (u_{\sigma,i}^n - u_{K,i}^n) \, \boldsymbol{j}^{(i)} \cdot \boldsymbol{n}_{K,\sigma},$$

where $\boldsymbol{j}^{(i)}$ stands for the ith vector of the canonical basis of $\mathbb{R}^d$ and $u_{K,i}^n$ stands for the mean value of the velocity over the two faces of $\mathcal{E}^{(i)}(K)$. In both cases, we obtain that:

$$\left| \sum_{\sigma \in \mathcal{E}(K)} |\sigma| \, u_{K,\sigma}^n \right| \le 2 \sum_{i=1}^d \sum_{(\sigma,\sigma') \in \mathcal{E}^{(i)}(K)^2} (|\sigma| + |\sigma'|) \, |u_{\sigma,i}^n - u_{\sigma',i}^n|.$$

Therefore,

$$|K| \, |(R_{01})_K^{n+1}| \le 2 \, |\varphi''|_\infty M \, |\rho_K^{n+1} - \rho_K^n|$$

$$\sum_{i=1}^d \sum_{(\sigma,\sigma') \in \mathcal{E}^{(i)}(K)^2} (|\sigma| + |\sigma'|) \, |u_{\sigma,i}^n - u_{\sigma',i}^n|.$$

We thus have, thanks to a Hölder estimate, for $q \ge 1$, $q' \ge 1$ and $\dfrac{1}{q} + \dfrac{1}{q'} = 1$:

$$\|R_{01}\|_{L^1} = \sum_{n=0}^{N-1} \delta t \sum_{K \in \mathcal{M}} |K| \, (R_{01})_K^{n+1}$$

$$\le 2 \, |\varphi''|_\infty M \left[\delta t \sum_{n=0}^{N-1} \sum_{K \in \mathcal{M}} |K| \, |\rho_K^{n+1} - \rho_K^n|^q \Big(\sum_{i=1}^d \sum_{(\sigma,\sigma') \in \mathcal{E}^{(i)}(K)^2} 1 \Big) \right]^{1/q}$$

$$\left[\sum_{n=0}^{N-1} \sum_{K \in \mathcal{M}} \sum_{i=1}^d \sum_{(\sigma,\sigma') \in \mathcal{E}^{(i)}(K)^2} \delta t \, |K| \left(\frac{|u_{\sigma,i}^n - u_{\sigma',i}^n|}{h_K} \right)^{q'} \left(\frac{(|\sigma| + |\sigma'|) \, h_K}{|K|} \right)^{q'} \right]^{1/q'}.$$

Using $|\rho_K^{n+1} - \rho_K^n|^q \le (2M)^{q-1} \, |\rho_K^{n+1} - \rho_K^n|$ yields (65).

The object of the following proposition is to mimick at the fully discrete level the computation of (2) and (3), applying it to $z = e$.

Proposition 2 (Inequalities Derived from the Internal Energy Balance) *Let* $M \geq 1$ *and let us suppose that* $\rho_K^n \leq M$, $e_K^n < M$, $1/e_K^n \leq M$ *and* $|u_{K,\sigma}| \leq M$, *for* $K \in \mathcal{M}$, $\sigma \in \mathcal{E}(K)$ *and* $0 \leq n \leq N$. *Assume that the face approximation of the internal energy satisfies (45). Let* φ *be a twice continuously convex function from* $(0, +\infty)$ *to* $\mathbb{R}$. *Then*

$$\varphi'(e_K^{n+1}) \left[\frac{|K|}{\delta t} (\rho_K^{n+1} e_K^{n+1} - \rho_K^n e_K^n) + \sum_{\sigma \in \mathcal{E}(K)} F_{K,\sigma}^n e_\sigma^n \right]$$

$$\geq \frac{|K|}{\delta t} (\rho_K^{n+1} \varphi(e_K^{n+1}) - \rho_K^n \varphi(e_K^n)) + \sum_{\sigma \in \mathcal{E}(K)} F_{K,\sigma}^n \varphi(e_\sigma^n) + |K| (R_e)_K^n, \tag{66}$$

with

$$\|R_e\|_{L^1} \leq 3M^2 |\varphi'|_\infty \|e\|_{\mathcal{T},x,\mathrm{BV}} h_\mathcal{M} + M^2 |\varphi''|_\infty \|e\|_{\mathcal{T},t,\mathrm{BV}} \frac{\delta t}{h_\mathcal{M}}, \tag{67}$$

where $|\varphi'|_\infty = \max(|\varphi'(1/M)|, |\varphi'(M)|)$, $|\varphi''|_\infty$ *stands for the maximum value taken by* φ'' *over the interval* $[1/M, M]$, *and* $h_\mathcal{M}$ *is defined by (23). If, furthermore, the approximation of* e_σ^n *in (66) is upwind, and the CFL condition (50) holds (with* φ *instead of* φ_e*) then* $(R_e)_K^n = 0$.

Proof First, the fully discrete identity corresponding to the semi-discrete identity (5) with $z = e$ is obtained thanks to the discrete mass equation; it reads:

$$\frac{|K|}{\delta t} (\rho_K^{n+1} e_K^{n+1} - \rho_K^n e_K^n) + \sum_{\sigma \in \mathcal{E}(K)} F_{K,\sigma}^n e_\sigma^n = \frac{|K|}{\delta t} \rho_K^{n+1} (e_K^{n+1} - e_K^n)$$

$$+ \sum_{\sigma \in \mathcal{E}(K)} F_{K,\sigma}^n (e_\sigma^n - e_K^n), \qquad \forall K \in \mathcal{M}, \quad 0 \leq n \leq N - 1.$$

Now let φ be a twice continuously differentiable function from $(0, +\infty)$ to $\mathbb{R}$, and let us multiply the first two terms of the discrete internal energy balance (44b) by $\varphi'(e_K^{n+1})$; switching from the conservative to the non conservative form, we get:

$$\varphi'(e_K^{n+1}) \left[\frac{|K|}{\delta t} (\rho_K^{n+1} e_K^{n+1} - \rho_K^n e_K^n) + \sum_{\sigma \in \mathcal{E}(K)} F_{K,\sigma}^n (e_\sigma^n - e_K^n) \right]$$

$$= (T_1)_K^{n+1} + (T_2)_K^{n+1} + |K| R_K^{n+1},$$

with

$$(T_1)_K^{n+1} = \varphi'(e_K^{n+1}) \frac{|K|}{\delta t} \rho_K^{n+1} (e_K^{n+1} - e_K^n),$$

$$(T_2)_K^{n+1} = \varphi'(e_K^n) \sum_{\sigma \in \mathcal{E}(K)} F_{K,\sigma}^n (e_\sigma^n - e_K^n),$$

$$|K| R_K^{n+1} = \big(\varphi'(e_K^{n+1}) - \varphi'(e_K^n)\big) \sum_{\sigma \in \mathcal{E}(K)} F_{K,\sigma}^n (e_\sigma^n - e_K^n). \tag{68}$$

The remainder term R_K^{n+1} is quite similar to the remainder defined by (57) in the proof of Proposition 1; following the proof of (60), we get that it satisfies:

$$\|R\|_{L^1} \le M^2 \, |\varphi''|_\infty \, \|e\|_{\mathcal{T},t,\mathrm{BV}} \, \frac{\delta t}{h_{\mathcal{M}}}.$$

Now

$$(T_1)_K^{n+1} = \frac{|K|}{\delta t} \rho_K^{n+1} \big(\varphi(e_K^{n+1}) - \varphi(e_K^n)\big) + |K| \, (R_1)_K^{n+1},$$

$$(T_2)_K^{n+1} = \sum_{\sigma \in \mathcal{E}(K)} F_{K,\sigma}^n \big(\varphi(e_\sigma^n) - \varphi(e_K^n)\big) + |K| \, (R_2)_K^{n+1},$$

with:

$$|K| \, (R_1)_K^{n+1} = \frac{|K|}{\delta t} \rho_K^{n+1} \big(\varphi(e_K^n) - \varphi(e_K^{n+1}) - \varphi'(e_K^{n+1})(e_K^n - e_K^{n+1})\big),$$

$$|K| \, (R_2)_K^{n+1} = \sum_{\sigma \in \mathcal{E}(K)} F_{K,\sigma}^n \big(\varphi(e_K^n) + \varphi'(e_K^n)(e_\sigma^n - e_K^n) - \varphi(e_\sigma^n)\big).$$

The remainder $(R_1)_K^{n+1}$ may be written:

$$(R_1)_K^{n+1} = \frac{1}{2\delta t} \rho_K^{n+1} \, \varphi''(e_K^{n+1/2}) \, (e_K^{n+1} - e_K^n)^2, \tag{69}$$

where $e_K^{n+1/2} \in [\![e_K^n, \, e_K^{n+1}]\!]$. Since φ is supposed to be convex, this term is non-negative. Let e_{KL}^n be the real number defined by Eq. (26) (and denoted in this latter relation by x_{KL}) with $x_K = e_K^n$ and $x_L = e_L^n$. Thanks to (45), by a computation similar to the implicit case, the remainder R_2 satisfies

$$R_2 \ge \delta R_2 \quad \text{and} \quad \|\delta R_2\|_{L^1} \le 3M^2 \, |\varphi'|_\infty \, \|e\|_{\mathcal{T},x,\mathrm{BV}} \, h_{\mathcal{M}}. \tag{70}$$

Switching back from the non-conservative formulation to the conservative formulation yields:

$$\varphi'(e_K^{n+1}) \frac{|K|}{\delta t} \rho_K^{n+1}(e_K^{n+1} - e_K^n) + \varphi'(e_K^n)\left[\sum_{\sigma \in \mathcal{E}(K)} F_{K,\sigma}^n (e_\sigma^n - e_K^n)\right] \geq$$

$$\frac{|K|}{\delta t}\left(\rho_K^{n+1}\varphi(e_K^{n+1}) - \rho_K^n \, \varphi(e_K^n)\right) + \sum_{\sigma \in \mathcal{E}(K)} F_{K,\sigma}^n \, \varphi(e_\sigma^n) + |K|(R_2)_K^{n+1}, \qquad (71)$$

which, thanks to (70), leads to (66) and (67).

Let us now suppose that the discretization of the internal energy convection term is upwind. In this case, we obtain for $(R_2)_K^{n+1}$:

$$|K| \, (R_2)_K^{n+1} = \frac{1}{2} \sum_{\sigma \in \mathcal{E}(K)} (F_{K,\sigma}^n)^- \, \varphi''(e_{K,\sigma}^n)(e_K^n - e_L^n)^2, \qquad (72)$$

where $e_{K,\sigma}^n \in [\![e_K^n, \, e_L^n]\!]$. The remainder R_K^{n+1} yields in the upwind case:

$$|K| \, R_K^{n+1} = -\varphi''(\tilde{e}_K^{n+1/2}) \, (e_K^{n+1} - e_K^n)\left[\sum_{\sigma \in \mathcal{E}(K)} (F_{K,\sigma}^n)^- (e_L^n - e_K^n)\right],$$

where $e_K^{n+1/2} \in [\![e_K^n, \, e_K^{n+1}]\!]$. So, thanks to the Young inequality:

$$|K||R_K^{n+1}| \leq \frac{1}{2} \sum_{\sigma \in \mathcal{E}(K)} (F_{K,\sigma}^n)^- \, \varphi''(e_{K,\sigma}^n)(e_L^n - e_K^n)^2$$

$$+ \frac{1}{2} \, (e_K^{n+1} - e_K^n)^2 \sum_{\sigma \in \mathcal{E}(K)} (F_{K,\sigma}^n)^- \frac{\varphi''(\tilde{e}_K^{n+1/2})^2}{\varphi''(e_{K,\sigma}^n)}.$$

In view of the expressions (69) and (72) of $(R_1)_K^{n+1}$ and $(R_2)_K^{n+1}$ respectively, we obtain that $(R_1)_K^{n+1} + (R_2)_K^{n+1} + R_K^{n+1} \geq 0$ thanks to the CFL condition (50), which yields the result.

Theorem 3 deserves the following comments:

- First, in the explicit case, we are able to prove neither a local nor a global discrete entropy inequality; we only obtain some weak inequalities that allow to show the consistency of the scheme, under some conditions.
- The convergence to zero with the space and time step of the remainders is obtained, supposing a control of discrete solutions in L^∞ and discrete BV norms, in two cases: first when the ratio $\delta t / \underline{h}_\mathcal{M}$ tends to zero, second when the $L^q(0, T; W_\mathcal{M}^{1,q})$ norm of the velocity does not blow-up too quickly with the

space step. To this respect, let us suppose that we implement a stabilization term in the momentum balance equation reading (in a pseudo-continuous setting, for short and to avoid the technicalities associated to the space discretization), for $1 \leq i \leq d$:

$$\partial_t(\rho u_i) + \mathrm{div}(\rho u_i \boldsymbol{u}) + \partial_i p - h_{\mathcal{M}}^{\alpha} \Delta_q u_i = 0, \tag{73}$$

where $\Delta_q u_i$ is such that

$$\|u_i\|_{W_{\mathcal{M}}^{1,q}}^q \leq C \int_{\Omega} -\Delta_q u_i \, u_i \, \mathrm{d}\boldsymbol{x},$$

where C is independent of $h_{\mathcal{M}}$. This kind of viscosity term may be found in turbulence models [1, 36]. Multiplying (73) by u_i and integrating with respect to space and time yields:

$$\int_0^T \int_{\Omega} -\Delta_q u_i \, u_i \, \mathrm{d}\boldsymbol{x} \, \mathrm{d}t = -\int_0^T \int_{\Omega} \left(\partial_t(\rho u_i) + \mathrm{div}(\rho u_i \boldsymbol{u}) + \partial_i p \right) u_i \, \mathrm{d}\boldsymbol{x} \, \mathrm{d}t. \tag{74}$$

In this relation, the right-hand side may be controlled under L^{∞} and BV stability assumptions (remember that, at the discrete level, the BV and $W^{1,1}$ norms are the same), and we obtain an estimate on $\|\boldsymbol{u}\|_{L^q(0,T;W_{\mathcal{M}}^{1,q})}$ which may be used in (55). A standard first order diffusion-like stabilizing term corresponds to $q = 2$ and $\alpha = 1$; it yields a bound on $h_{\mathcal{M}}^{1/2} \|\boldsymbol{u}\|_{L^2(0,T;H_{\mathcal{M}}^1)}$, so that (55) becomes

$$\|R_\eta\|_{L^1} \leq f_{\mathcal{M}} \, C_{\mathcal{M}} \, M^{\frac{3}{2}} \, |\varphi''|_{\infty} \, \|\rho\|_{T,t,\mathrm{BV}}^{\frac{1}{2}} \, \tilde{C}(\frac{\delta t}{h_{\mathcal{M}}})^{\frac{1}{2}}.$$

Such a stabilization is thus not sufficient to ensure that the remainder term tends to zero. What is needed is in fact:

$$\alpha < q - 1.$$

To avoid an over-diffusion in the momentum balance, this inequality suggests to implement a non-linear stabilization with $q > 2$ which, in turn, will allow $\alpha > 1$. With such a trick, we will be able to obtain for first-order upwind schemes the desired "Lax-consistency" result: the limit of a convergent sequence of solutions, bounded in L^{∞} and BV norms, and obtained with space and time steps tending to zero, satisfies a weak entropy inequality.

– We introduced in [33] a limitation process for a MUSCL-like algorithm for the transport equation, which consists in deriving an admissible interval for the approximation of the unknowns at the mesh faces, in convection terms, thanks to extrema preservation arguments. This limitation process has been extended to the

Euler equations in [14]. The conditions (45) and (46) may easily be incorporated in this limitation: indeed, they also define an admissible interval, which is not disjoint from the MUSCL-like admissible interval of [33], since the upwind value belongs to both. A similar idea (namely restricting the choice for the face approximation in order to obtain an entropy inequality) may be found in [2].

References

1. Berselli, L., Illiescu, T., Layton, W.: Mathematics of Large Eddy Simulation of Turbulent Flows. Springer, New York (2006)
2. Berthon, C., Desveaux, V.: An entropy preserving MOOD scheme for the Euler equations. Int. J. Finite Vol. **11**, 1–39 (2014)
3. Bouchut, F.: Nonlinear stability of finite volume methods for hyperbolic conservation laws and well-balanced schemes for sources. In: Frontiers in Mathematics. Birkhäuser Verlag, Basel (2004)
4. CALIF^{3}S: A software components library for the computation of reactive turbulent flows. https://gforge.irsn.fr/gf/project/isis
5. Chiodaroli, E., Feireisl, E., Kreml, O.: On the weak solutions to the equations of a compressible heat conducting gas. Annales de l'Institut Henri Poincaré. Analyse Non Linéaire **32**, 225–243 (2015)
6. Ciarlet, P.G.: Basic error estimates for elliptic problems. In: Ciarlet, P., Lions, J. (eds.) Handbook of Numerical Analysis, vol. II, pp. 17–351. North Holland, Amsterdam (1991)
7. Coquel, F., Helluy, P., Schneider, J.: Second-order entropy diminishing scheme for the Euler equations. Int. J. Numer. Meth. Fluids **50**, 1029–1061 (2006)
8. Crouzeix, M., Raviart, P.: Conforming and nonconforming finite element methods for solving the stationary Stokes equations. RAIRO Série Rouge **7**, 33–75 (1973)
9. Dakin, G., Després, B., Jaouen, S.: High-order staggered schemes for compressible hydrodynamics. Weak consistency and numerical validation. J. Comput. Phys. **376**, 339–364 (2019)
10. Eymard, R., Gallouët, T., Herbin, R.: Finite volume methods. In: Ciarlet, P., Lions, J. (eds.) Handbook of Numerical Analysis, vol. VII, pp. 713–1020. North Holland, Amsterdam (2000)
11. Feireisl, E., Hošek, R., Michálek, M.: A convergent numerical method for the full Navier-Stokes-Fourier system in smooth physical domains. SIAM J. Numer. Anal. **54**, 3062–3082 (2016)
12. Gallouët, T., Gastaldo, L., Herbin, R., Latché, J.C.: An unconditionally stable pressure correction scheme for compressible barotropic Navier-Stokes equations. Math. Modell. Numer. Anal. **42**, 303–331 (2008)
13. Gallouët, T., Herbin, R., Latché, J.C.: Kinetic energy control in explicit finite volume discretizations of the incompressible and compressible Navier-Stokes equations. Int. J. Finite Vol. **7**(2), 1–6 (2010)
14. Gastaldo, L., Herbin, R., Latché, J.C., Therme, N.: A MUSCL-type segregated - explicit staggered scheme for the Euler equations. Comput. Fluids **175**, 91–110 (2018)
15. Godunov, S.K.: A difference method for numerical calculation of discontinuous solutions of the equations of hydrodynamics. Mat. Sb. (N.S.) **47**(89), 271–306 (1959)
16. Goudon, T., Llobell, J., Minjeaud, S.: A staggered scheme for the Euler equations. In: Finite Volumes for Complex Applications VIII - Problems and Perspectives - Lille (2017)

17. Grapsas, D., Herbin, R., Kheriji, W., Latché, J.C.: An unconditionally stable staggered pressure correction scheme for the compressible Navier-Stokes equations. SMAI-J. Comput. Math. **2**, 51–97 (2016)
18. Guillard, H.: Recent developments in the computation of compressible low Mach flows. Flow Turbul. Combust. **76**, 363–369 (2006)
19. Harlow, F., Amsden, A.: A numerical fluid dynamics calculation method for all flow speeds. J. Comput. Phys. **8**, 197–213 (1971)
20. Harlow, F., Welsh, J.: Numerical calculation of time-dependent viscous incompressible flow of fluid with free surface. Phys. Fluids **8**, 2182–2189 (1965)
21. Herbin, R., Kheriji, W., Latché, J.C.: On some implicit and semi-implicit staggered schemes for the shallow water and Euler equations. Math. Modell. Numer. Anal. **48**, 1807–1857 (2014)
22. Herbin, R., Latché, J.C., Nguyen, T.: Consistent segregated staggered schemes with explicit steps for the isentropic and full Euler equations. Math. Modell. Numer. Anal. **52**, 893–944 (2018)
23. Herbin, R., Latché, J.C., Minjeaud, S., Therme, N.: Conservativity and weak consistency of a class of staggered finite volume methods for the Euler equations. Math. Comput. (2020) https://doi.org/10.1090/mcom/3575
24. Herbin, R., Latché, J.C., Saleh, K.: Low mach number limit of some staggered schemes for compressible barotropic flows (2019, submitted). https://arxiv.org/abs/1803.09568
25. Herbin, R., Latché, J.C., Zaza, C.: A cell-centered pressure-correction scheme for the compressible Euler equations. IMAJNA. J. Numer. Anal. **40**(3), 1792–1837 (2020)
26. Ismail, F., Roe, P.L.: Affordable, entropy-consistent Euler flux functions. II. Entropy production at shocks. J. Comput. Phys. **228**, 5410–5436 (2009)
27. Larrouturou, B.: How to preserve the mass fractions positivity when computing compressible multi-component flows. J. Comput. Phys. **95**, 59–84 (1991)
28. Latché, J.C., Saleh, K.: A convergent staggered scheme for variable density incompressible Navier-Stokes equations. Math. Comput. **87**, 581–632 (2018)
29. Liou, M.S.: A sequel to AUSM, part II: AUSM+-up. J. Comput. Phys. **214**, 137–170 (2006)
30. Liou, M.S., Steffen, C.: A new flux splitting scheme. J. Computat. Phys. **107**, 23–39 (1993)
31. Llobell, J.: Schémas volumes finis à mailles décalées pour la dynamique des gaz. Ph.D. thesis, Université Côte d'Azur (2018)
32. Mardane, A., Fjordholm, U., Mishra, S., Tadmor, E.: Entropy conservative and entropy stable finite volume schemes for multi-dimensional conservation laws on unstructured meshes. In: European Congress Computational Methods Applied Sciences and Engineering, Proceedings of ECCOMAS 2012, held in Vienna (2012)
33. Piar, L., Babik, F., Herbin, R., Latché, J.C.: A formally second order cell centered scheme for convection-diffusion equations on general grids. Int. J. Numer. Meth. Fluids **71**, 873–890 (2013)
34. Rannacher, R., Turek, S.: Simple nonconforming quadrilateral Stokes element. Numer. Meth. Part. Diff. Equ. **8**, 97–111 (1992)
35. Ray, D., Chandrashekar, P., Fjordholm, U.S., Mishra, S.: Entropy stable scheme on two-dimensional unstructured grids for Euler equations. Commun. Comput. Phys. **19**(5), 1111–1140 (2016)
36. Sagaut, P.: Large Eddy Simulation for Incompressible Flows: An Introduction. Springer, New York (2006)
37. Steger, J., Warming, R.: Flux vector splitting of the inviscid gaz dynamics equations with applications to finite difference methods. J. Comput. Phys. **40**, 263–293 (1981)
38. Tadmor, E.: Entropy stable schemes. In: Abgrall, R., Shu, C.W. (eds.) Handbook of Numerical Analysis, vol. XVII, pp. 767–493. North Holland, Amsterdam (2016)
39. Toro, E.: Riemann Solvers and Numerical Methods for Fluid Dynamics – A Practical Introduction, 3rd edn. Springer, New York (2009)
40. Toro, E., Vázquez-Cendón, M.: Flux splitting schemes for the Euler equations. Comput. Fluids **70**, 1–12 (2012)

41. Van Leer, B.: Towards the ultimate conservative difference scheme. V. A second-order sequel to Godunov's method. J. Comput. Phys. **32**, 101–136 (1979)
42. Wesseling, P.: Principles of Computational Fluid Dynamics. Springer Series in Computational Mathematics, vol. 29. Springer, New York (2001)
43. Zha, G.C., Bilgen, E.: Numerical solution of Euler equations by a new flux vector splitting scheme. Int. J. Numer. Meth. Fluids **17**, 115–144 (1993)

Comparison and Analysis of Natural Laminar Flow Airfoil Shape Optimization Results at Transonic Regime with Bumps and Trailing Edge Devices Solved by Pareto Games and EAs

Yongbin Chen, Zhili Tang, and Jacques Periaux

Abstract The transonic natural laminar flow wing will become an important feature of the next generation advanced civil transport aircraft, because it can greatly reduce the friction drag. In paper Tang et al. (Arch Computat Meth Eng 26:119–141, 2019) and Chen et al. (J Nanjing Univ Aeronaut Astron, 50(4):548–557, 2018), the problem of wave drag increase due to the expansion of laminar flow region in the optimization design of natural laminar airfoil is studied with Pareto game and EAs by using Shock Control Bump (SCB) and Trailing Edge Device (TED) respectively. In this paper, the numerical implementation of SCB and TED in the shape design optimization of natural laminar airfoils and the performance differences of the final optimal airfoil are compared and analyzed. The feasibility and potential of applying them to the optimization design of three-dimensional laminar wing are discussed.

Keywords Natural laminar airfoil · Shock control bump · Trailing edge device · Multi-objective optimization · Evolutionary optimization · Pareto games

Y. Chen
College of Aerospace Engineering, Nanjing University of Aeronautics and Astronautics, Nanjing, China

Institute of Aeronautics Engineering and Technology, Binzhou University, Binzhou, China

Z. Tang (✉)
College of Aerospace Engineering, Nanjing University of Aeronautics and Astronautics, Nanjing, China
e-mail: tangzhili@nuaa.edu.cn

J. Periaux
International Center for Numerical Methods in Engineering (CIMNE), Barcelona, Spain

University of Jyvaskyla, IT Faculty, Jyväskylä, Finland

© The Author(s), under exclusive license to Springer Nature Switzerland AG 2021
D. Greiner et al. (eds.), *Numerical Simulation in Physics and Engineering: Trends and Applications*, SEMA SIMAI Springer Series 24,
https://doi.org/10.1007/978-3-030-62543-6_4

155

1   Introduction

In order to improve the performances of a civil transport aircraft at transonic regimes, it is critical to develop new computational optimization methods reducing friction drag [1], for example, laminarity technology. In paper [2], the feasibility of NLF airfoil/wing shape design optimization in transonic regime is investigated, i.e. delaying the transition location to maintain a larger region of favorable pressure gradient on airfoil surface, meanwhile installing an optimal SCB shape at the location of shock wave [3] or an optimal TED shape at the trailing edge [4] to reduce wave drag. Therefore the single natural laminar flow (NLF) airfoil design optimization at transonic regimes is converted into the following two-objective optimization problem as presented in (1).

$$\begin{cases} \max_{(Airfoil,SCB/TED)} \mathcal{J}_1 = x_{upper} + x_{lower} \\ \min_{(Airfoil,SCB/TED)} \mathcal{J}_2 = C_{D_{wave}} \end{cases} \tag{1}$$

where x_{upper} and x_{lower} are transition locations on upper and lower surfaces of an airfoil respectively, $C_{D_{wave}}$ is the wave drag, and SCB/TED denotes design variables of shape/position of a Bump device or of a Trailing Edge device respectively.

The research numerical methods and results of natural laminar flow airfoil optimization using SCB or TED devices have been given in details in [3] and [4], respectively, using Pareto game and EAs. Numerical experiments demonstrate that Pareto game coupled to the EAs optimizer can easily capture a Pareto front, a set of solutions of this two-objective shape optimization problem. This paper compare numerical results obtained by a Pareto game formulation considering SCB or TED devices. They include:

- Drag reduction performances of airfoil trade offs solutions between the delay of the profiles transition location and the increased intensity of shock wave varying with the position and shape of a bump installed on the upper surface of the airfoil or with the effect of trailing edge device;
- Computational efficiency versus quality design of the Pareto non dominated solutions of the hybridized Game-EAs software in the cases of SCB or TED devices actions;
- From the analysis of 2D results, it is concluded that a variety of laminar flow airfoils with greener aerodynamic performances (drag reduction) can be significantly improved due to either optimal SCB shape and position or TED shape when compared to the baseline airfoil geometry. It is concluded that this new coupling relaminarization-active device methodology confirm the potential of such an approach to solve the challenging 3-D optimization of the natural laminar flow of natural laminar wings in industrial design environments.

2 NLF Airfoil Shape Design Optimization with Bump and TED Devices for Shock Wave Control

A GAs hybridized with a Pareto game is implemented to optimize the airfoil shape with a larger laminar flow range and a weaker shock wave drag simultaneously due to the action of a SCB [3] or a TED [4], as presented in (1). The RAE2822 airfoil is the baseline airfoil shape. The second order numerical scheme Reynolds averaged Navier-Stokes (RANS) simulation coupled with a boundary layer analysis [5, 6] and a linear stability prediction method e^N is introduced into the NLF airfoil design optimization with only one flow analysis needed and since only one flow analysis and less time-consuming sub-iterations between boundary layer analysis and transition prediction is required [7]. The value $N = 9$ is chosen as the threshold value according to the engineering experience. Then, the procedure is the following: substitute the transition location into the boundary layer equations, solve the boundary layer equations and Orr-Sommerfeld equation, and repeats this procedure until the transition location does not change any more.

The airfoil shape is parameterized by using Bézier-curve, as shown in Fig. 1, and $y_{1,up}, \cdots, y_{7,up}$ and $y_{1,low}, \cdots, y_{7,low}$ are coordinates of control points of Bézier curves. The SCB shape is defined by a Hicks-Henne function, where x_{heigth}, x_{length} and $x_{relative}$ represent design parameters for SCB, as shown in Fig. 2. The corresponding search spaces for airfoil and SCB optimization are shown on Tables 1 and 2 respectively.

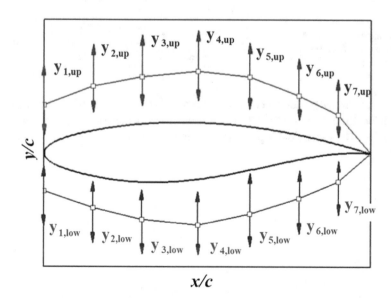

Fig. 1 The sketch diagram of airfoil parameterization. From Tang et al. [3]. Reproduced with the permission of Springer Publishing

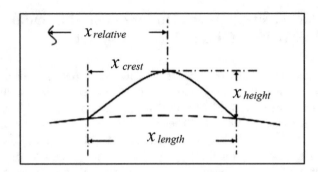

Fig. 2 The sketch diagram of the bump parameterization. From Tang et al. [3]. Reproduced with the permission of Springer Publishing

Table 1 The search space of airfoil shape (c is the chord length of an airfoil). From Tang et al. [3]

Parameters	Lower bound	Upper bound	Parameters	Lower bound	Upper bound
$y_{1,up}/c$	−0.002	0.002	$y_{1,low}/c$	−0.002	0.002
$y_{2,up}/c$	−0.003	0.003	$y_{2,low}/c$	−0.003	0.003
$y_{3,up}/c$	−0.005	0.005	$y_{3,low}/c$	−0.005	0.005
$y_{4,up}/c$	−0.005	0.005	$y_{4,low}/c$	−0.005	0.005
$y_{5,up}/c$	−0.005	0.005	$y_{5,low}/c$	−0.005	0.005
$y_{6,up}/c$	−0.003	0.003	$y_{6,low}/c$	−0.003	0.003
$y_{7,up}/c$	−0.003	0.003	$y_{7,low}/c$	−0.002	0.002

Table 2 The search space of bump shape (c is the chord length of an airfoil). From Tang et al. [3]

	$x_{relative}/c$	x_{length}/c	x_{height}/c
Lower bound	−0.05	0.10	0.001
Upper bound	0.05	0.30	0.005

The trailing edge device could also be used to control the wave drag, and the sketch diagram of a cambered TED is shown in Fig. 3 which shape is fitted by Akima functions. The corresponding search spaces in optimization for TED is given in Table 3.

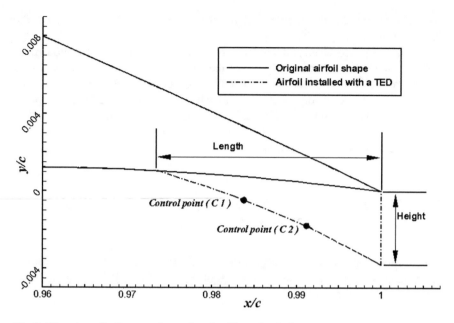

Fig. 3 The schematic diagram of a cambered trailing edge device

Table 3 The search space of trailing edge device shape (c is the chord length of an airfoil)

	x_{length}/c	x_{height}/c	$C1$	$C2$
Lower bound	0.2/c	0.002/c	$0.15x_{height}$	$0.5x_{height}$
Upper bound	0.3/c	0.005/c	$0.45x_{height}$	$0.7x_{height}$

3   Comparison and Analysis of Optimization Results Using SCB and TED

In this section, the theoretical and numerical tools defining the modeling of NLF airfoil optimization, parameterization, flow analysis and transition prediction of Sect. 2 are used for the implementation of the optimization procedure of NLF airfoils using active devices like bumps or Trailing Edge. In order to reduce the wave drag due to the existence of long favorable pressure gradient on the upper surface of NLF airfoil, a passive/active shock control device SCB and TED have to be installed on the airfoil surface at the location of shock wave or trailing edge in order to reduce its wave intensity. A parallelized version of a Non-dominated Sorting Genetic Algorithm II [8] (NSGA-II introduced by K. Deb) is used to solve the two-objective laminar airfoil shape optimization problem (1). The optimization results and numerical implementation are listed in detail in [3] and [4] respectively. The design flight conditions are: $M_\infty = 0.729$, $\alpha = 2.31°$, $Re = 1.28 \times 10^7$. In two NLF optimizations with bump or TED, the population size is 150, the crossover probability is 0.8, the mutation probability is 0.1, and the selection operator is the Pareto-tournament algorithm.

The convergence history of the two-objective NLF design optimization problem (1) using SCB is shown in Fig. 4, and the converged Pareto front achieved after 80 generations. The solutions of converged Pareto front are shown in Fig. 5.

The convergence history of the two-objective NLF design optimization problem (1) with TED is shown in Fig. 6. The Pareto front using this TED device is captured by evaluation of candidates of the population after 80 generations of the optimization procedure, as shown on Fig. 7.

3.1 Comparison and Analysis of Results: Advantages and Drawbacks of SCB and TED Devices

This section consists in the comparison of numerical results obtained with the Pareto game maximizing the laminar region of the airfoil and minimising the shock intensity with SCB or TED devices.

3.1.1 Comparing Efficiency of SCB Technology Versus TED Technology

In order to perform the two computations—one with Bump or the other with TED devices—it was necessary to check the similarity of the elements of simulation: they are compared on Table 4 in considering mesh size, design parameters, design variables, GAs parameters (population size, generations, cross over and mutation). Considered as only one test case computed with the two devices (SCB or TED) the flight conditions were identical, as well as the flow analyzer giving access to the location of the transition point. The NSGA-II optimization procedures for both simulations are also similar.

Analysing the results of the two computations, it is noticed that the captured Pareto front with the SCB technology and the one with the TED technology have a similar shape with a discontinuity region differently located on the interval [0.7, 0.9] of the airfoil chord for the SCB device and [0.82, 0.95] of the airfoil chord for the TED device. On the two Pareto fronts, it can be observed that the Pareto Member B (PM B) solution is located before the discontinuity region (0.68 and 0.8 respectively) while the PM C (0.97 and 0.98 respectively) and PM A (1.0 and 1.2 respectively) solutions are located after the discontinuity region.

All data needed to run the numerical simulation of the two NLF optimization problem using Bump or TED technologies are provided on Table 4. It is observed that the CPU cost for both simulations is quite similar due to the same number of design variables (17 for the SCB and 18 for the TED) and quite similar number

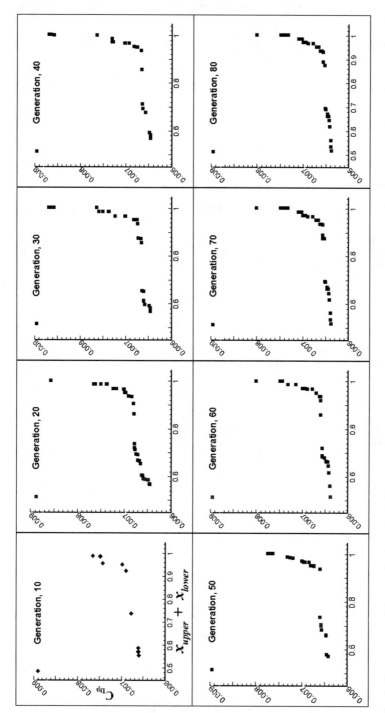

Fig. 4 Convergence of the non-dominated solutions at different generations of the two objective NLF airfoil shape optimization with bump. From Tang et al. [3]. Reproduced with the permission of Springer Publishing

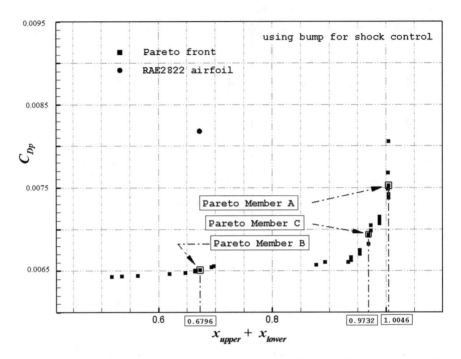

Fig. 5 Converged Pareto Front and the solution of RAE2822 airfoil. From Tang et al. [3]. Reproduced with the permission of Springer Publishing

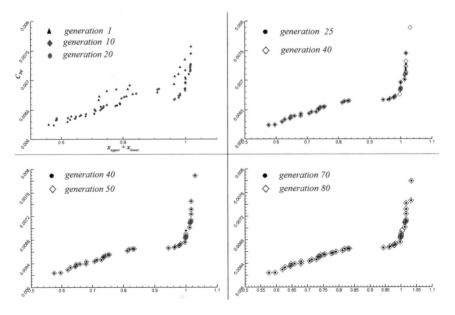

Fig. 6 The convergence history of two-objective NLF optimization with TED technology

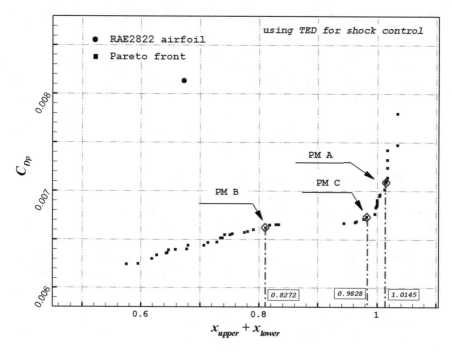

Fig. 7 The captured Pareto front of the Multi-objective problem with TED

of mesh points (50,000 for the SCB versus 55,000 for the TED). It is noticed that parameters for the GAs are the same in the to optimization procedures. Finally the two cases are run on the same computer E5-2640 with 256 cores.

Table 4 Pareto front with Bump and TED. From Tang et al. [3]. Reproduced with the permission of Springer Publishing

Pareto front	With bump	With TED
Mesh size	50,000	55,000
Number of parameters	17(14+3)	18(14+4)
Population size	150	150
Generations	80	80
Crossover	0.8	0.8
Mutation	0.1	0.1
Number of CPU cores	256	256
CPU performance	E5-2640	E5-2640
CPU cost (h)	140	140

3.1.2 Comparing the Design Quality Level of SCB Technology and TED Technology

It can be observed on Fig. 8 that maximizing the laminar region and minimizing the intensity of the shock simultaneously provide a significant design improvement of the RAE2822 airfoil baseline. On Fig. 8 are plotted that pressure distributions and transitions locations of the RAE2822 airfoil equipped with optimized bump shape or not. It can be easily observed how the transition point locations of PM A, PM B and PM C are delayed with the optimization and the shock intensity is smoothed and reduced on Figs. 5 and 7 (Fig. 9).

For design quality comparison are provided on Table 5 important aerodynamical data (transition locations at the upper and lower part of the airfoil and the value of C_{Dpre} replacing C_{Dwave}). A careful analysis of data provided on Table 5 suggests the following conclusions: both active device concepts improve globally the design quality (greener!) of the NLF airfoil and among the different Pareto members the PM C solution is the best compromised one for the bump technology while PM C is also the best compromised one for the TED technology. It remains the decision of the designer to choose tradeoff solutions among non-dominated trade off solutions for the design project. In addition, almost the same quality of optimal Pareto fronts are obtained by using SCB and TED devices respectively, as shown in Fig. 10.

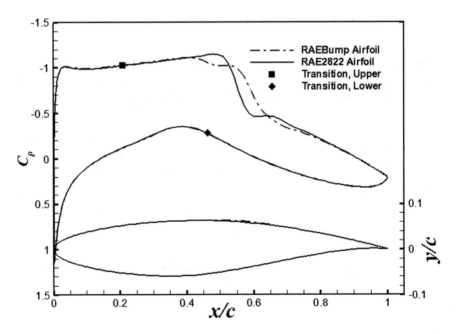

Fig. 8 Pressure distribution and transition location on RAE2822 airfoil and airfoil equipped with a bump. From Tang et al. [3]. Reproduced with the permission of Springer Publishing

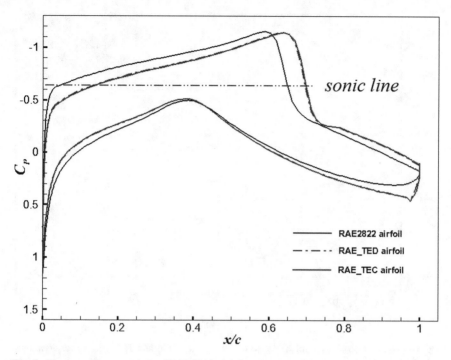

Fig. 9 The pressure comparison of RAE2822 airfoil and same airfoil with trailing edge devices

Table 5 The aerodynamic performance of airfoils with bump and TED technologies. From Tang et al. [3]. Reproduced with the permission of Springer Publishing

	C_L	C_{Dpre}	x_{upper}/c	x_{lower}/c
RAE2822	0.7064	0.008095	0.2102	0.4624
Data for airfoils with bump				
PM A	0.7181	0.007428	0.5416	0.4630
PM B	0.7016	0.006358	0.2166	0.4630
PM C	0.7056	0.006859	0.5104	0.4628
Data of airfoils with TED				
PM A	0.7067	0.007092	0.5558	0.4587
PM B	0.7065	0.006641	0.3687	0.4585
PM C	0.7066	0.006734	0.5220	0.4608

3.2 Extension to the 3-D NLF Wing Design Optimization with 3-D SCB/TED Devices

With the extension to 3D laminar NLF wing/aircraft design optimization, the shock control bump concept is numerically confronted to the problem of requiring simultaneous different grids around a series of bumps in order to simulate accurately the flow disturbance generated by bumps in the vicinity of the wing surface in chord and span wise directions. These 3-D bump geometries generate a series of local fine grids associated to each bump compared to the original baseline wing, shown in

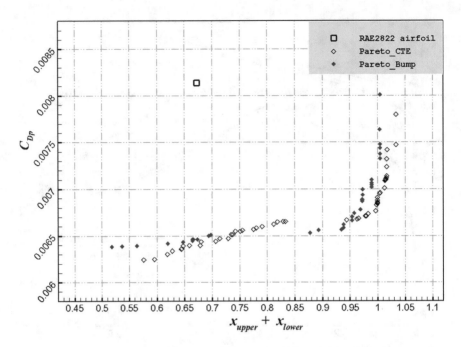

Fig. 10 Comparison of the optimal Pareto fronts by using SCB and TED devices for wave drag reduction

Fig. 11 The grid around original wing and the grid around wing installed with bump. From Chen [9]

Fig. 11. Consequently, additional grids linked to the series will increase significantly the CPU cost of the flow analyser (about five times on a E5 processor with 8 cores as presented on Table 6 when comparing grid sizes used for SCB technology versus grid sizes used for TED technology).

Based on the above remarks, the TED technology (see geometry details of trailing edge installed on a wing on Fig. 12) will be quite often preferred in industrial design environment to control and decrease the 3-D shock wave intensity. A two objective optimization of a 3-D Wing/Fuselage maximizing simultaneously the laminarity region with the TED concept can be found in [9]. The capture of the Pareto Front surface has been computed with similar optimisation and flow analysis tools used in 2-D NLF airfoils. Due to CPU limitations, this 3-D design is

Table 6 The CPU cost of different wing cells in flow field simulation. From Chen [9]

	Straight wing section	Straight wing section with bump	Straight wing section with TED
Computer performance		E5 processor/8 cores	
Turbulent model		S-A model	
Cells	270,000	980,000	380,000
CPU cost (Min)	9	46	16

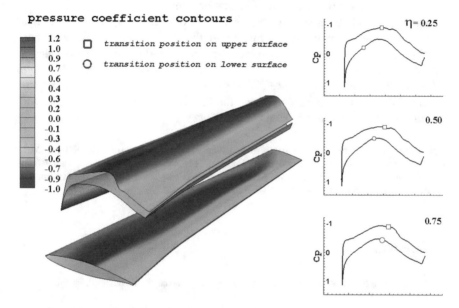

Fig. 12 The sketch diagram of wing installed with trailing edge device. From Chen [9]

accelerated in introducing hierarchical levels of physical flow modelings (potential flow, Euler flow and RANS flow). The pressure distribution on the wing of a non dominated solution of the Pareto front (refer to a figure of 3-D wing with TED, shown on Fig. 12) shows the positive effect of the TED device in reducing the shock intensity of the extrados of the wing while maximizing the laminarity surface region. These results show the potentiality of the approach for 3-D industrial design. The next step to investigate is the speed up the optimization procedure using TED technology in introducing a Nash-Pareto games coalition strategy to reduce the number of generations necessary to capture the set of non dominated optimized solutions. AI assistance with Machine Learning (Neural Networks) will be also investigated to decrease significantly the CPU cost in the evaluation of laminar wing candidates.

4 Conclusion and Perspectives

From the analysis of 2-D numerical results, it is concluded that a variety of laminar flow airfoils with greener aerodynamic performances can be significantly improved with the optimal shape and position of a SCB or TED devices compared to the baseline airfoil geometry. The role of active device in the NLF methodology confirms the potential of such an approach applied also to the 3-D shapes design optimization of NLF wings in industrial environments.

The capture of the non dominated solutions on the Pareto front of the two-objective optimization problem provides to the designer trade offs with non-dominated (Pareto) solutions. These compromised solution increase the delay of transition location and decrease the shock wave intensity simultaneously due to the bump installed on the surface (SCB) of the airfoil or the control trailing edge device (TED). Pareto solutions provide the best set of laminar flow airfoils with significantly improved aerodynamic performances.

Acknowledgments This research has also benefited of a partial support from EC and MIIT with the EU-China international cooperation project "MARS" [10]. The authors acknowledge also NUAA and CIMNE colleagues for fruitful discussions on game theory and their institutions providing partial support during crossed visits of the authors.

References

1. Quadrio, M., Ricco, P.: The laminar generalized stokes layer and turbulent drag reduction. J. Fluid Mech. **667**, 135–157 (2011)
2. Chen, Y.B., Tang, Z.L., Sheng, J.D.: Multi objective optimization for natural laminar flow airfoil in transonic flow. Chin. J Phys **33**(3), 283–296 (2016)
3. Tang, Z.L., Chen, Y.B., Zhang, L.H., Periaux, J.: Solving the two objective evolutionary shape optimization of a natural laminar airfoil and shock control Bump with Game Strategies. Arch. Computat. Meth. Eng. **26**, 119–141 (2019)
4. Chen, Y.B., Tang, Z.L., Sheng, J.D.: Trailing edge device application to wave drag reduction in NLF airfoil multi-island design optimization. J. Nanjing Univ. Aeronaut. Astron. **50**(4), 548–557 (2018)
5. Cebeci, T., Cousteix, J.: Modeling and Computation of Boundary Layer Flow: Laminar, Turbulent and Transitional Boundary Layers in Compressible Flows. Horizons Publishing, Long Beach (2003)
6. Cebeci, T., Keller, H.B.: Shooting and parallel shooting methods for solving the Falker-Skan boundary-layer equation. J. Comput. Phys. **7**, 289–290 (1971)
7. Langtry, R.B., Menter, F.R.: Correlation-based transition modeling for unstructured parallelized computational fluid dynamics codes. AIAA J. **47**(12), 2894–2906 (2009)
8. Deb, K., Agrawal, S., Pratap, A.: A fast and elitist multi-objective genetic algorithm: NSGA-II. IEEE T. Evolut. Comput. **6**, 182–197 (2002)
9. Chen, Y.B.: Natural Laminar Flow Shape Optimization Method and Its Applications in Wing Design at Transonic Regimes. College of Aerospace Engineer, Nanjing (2017)
10. Qin, N.: Advances in Effective Separation Control for Aircraft Applications, ECCOMAS. Springer Series on Computational Methods in Applied Sciences. Springer, New York (2019)

Time-Parallel Algorithm for Two Phase Flows Simulation

Katia Ait-Ameur, Yvon Maday, and Marc Tajchman

Abstract In this paper, we will report our recent effort to apply the parareal algorithm to the time parallelization of an industrial code that simulates two phase flows in a reactor for safety studies. This software solves the six equation two-fluid model by considering a set of balance laws (mass, momentum and energy) for each phase, liquid and vapor, of the fluid. The discretization is based on a finite volume method on a staggered grid in space and on a multistep time scheme. Here, we apply a variant of the parareal algorithm on an oscillating manometer test case: the multistep variant allowing to deal with multistep time schemes in the coarse and/or fine propagators. Numerical results show that parareal methods offer the potential for an increased level of parallelism and is a good strategy to complement the current space domain decomposition implemented in the code.

Keywords Time domain decomposition · Two phase flows · Parareal algorithm

K. Ait-Ameur (✉)
Laboratoire Jacques Louis Lions (LJLL), Sorbonne Université, Paris, France

CEA Saclay - CEA-DES/ISAS/DM2S/STMF/LMES, Gif-Sur-Yvette Cedex, France

Y. Maday
Laboratoire Jacques Louis Lions (LJLL), Sorbonne Université, Paris, France

Institut Universitaire de France, Maison des Universités, Paris, France
e-mail: maday@ann.jussieu.fr

M. Tajchman
CEA Saclay - CEA-DES/ISAS/DM2S/STMF/LMES, Gif-Sur-Yvette Cedex, France
e-mail: marc.tajchman@cea.fr

© The Author(s), under exclusive license to Springer Nature Switzerland AG 2021
D. Greiner et al. (eds.), *Numerical Simulation in Physics and Engineering: Trends and Applications*, SEMA SIMAI Springer Series 24,
https://doi.org/10.1007/978-3-030-62543-6_5

1 Introduction

In the nuclear energy domain, computations of complex two phase flows are required for the design and safety studies of nuclear reactors. System codes are dedicated to the thermalhydraulic analysis of nuclear reactors at the system scale by simulating the whole reactor. We are here interested in the Cathare code developed by the CEA. Like all system codes, Cathare essentially simulates assemblies of one-dimensional elements (pipes) and 3D elements (vessels). The discretization level of each element is kept intentionally at a coarse level to be able to handle whole systems simulations. Typical meshes used for the simulations are about 10^2 to 10^3 cells with a 3D element. Typical cases involve up to a million of numerical time iterations, computing the approximate solution during long physical simulation times. A space domain decomposition method has already been implemented and to improve the response time, we will consider a strategy of time domain decomposition, based on the *parareal method* [13]. The paper is organized as follows: after the presentation of the six equation two-fluid model and the Cathare numerical scheme in Sect. 2, the main aspects of the parareal methods will be recalled in Sect. 3. The numerical convergence observed in our example is shown in Sect. 4 followed by the performances we obtain by applying the multistep variant of the parareal algorithm.

2 Model

At the system scale, the finest details of the flow (description of the liquid-vapor interfaces for example) are not absolutely necessary to obtain a satisfactory macroscale description of the dynamics. For this reason macroscopic models have been developed that focus on the evolution of averaged quantities (see [12] and [7]). There are many different averaged models depending on the simplifying assumptions. The model used in Cathare is the 6 equation two-fluid model that considers a set of balance laws (mass, momentum and energy) for each phase, liquid and vapor. It assumes independent velocities and a pressure equality.

The unknowns are the volume fraction $\alpha_k \in [0, 1]$, the pressure $p \geq 0$, the velocity u_k and the enthalpy H_k of each phase. The subscript k stands for l if it is the liquid phase and g for the gas phase. For the sake of simplicity, we write the terms of the model involved in our test case, studied in Sect. 4.

$$
\begin{cases}
\partial_t (\alpha_k \rho_k) + \partial_x (\alpha_k \rho_k u_k) = 0 \\[2mm]
\alpha_k \rho_k \partial_t u_k + \alpha_k \rho_k u_k \partial_x u_k + \alpha_k \partial_x p = \alpha_k \rho_k g + F_k^{\text{int}} \\[2mm]
\partial_t \left[\alpha_k \rho_k \left(H_k + \frac{u_k^2}{2} \right) \right] + \partial_x \left[\alpha_k \rho_k u_k \left(H_k + \frac{u_k^2}{2} \right) \right] = \alpha_k \partial_t p + \alpha_k \rho_k u_k g
\end{cases}
$$

$$(1)$$

with $\alpha_g + \alpha_l = 1$ and the two equations of state: $\rho_k = \rho_k(p, H_k)$.

The interfacial forces F_k^{int} are of two types. The first ensures hyperbolicity of the system (see [15] for the well-posedness of the 6 equation model). The second is the interfacial friction term which will be important in the sequel for our test case. In this configuration, the phases are separated which means that one of the two phases vanishes in some parts of the domain. It is numerically challenging to compute the velocity of the ghost phase (see [16]). For this reason, the Cathare scheme forces the two velocities to be equal with this damping term.

2.1 Numerical Method

The Cathare scheme is based on a finite volume method on a staggered grid (MAC-type scheme) and on a two step time scheme. In a staggered scheme the i-th component of the velocity is located at the center of the edge orthogonal to the i-th unit vector. Pressures, void fractions and enthalpies are cell-centered. Given a time discretization of the full time interval $[0, T]$, we use the following notations: $(\alpha_k \rho_k)^n$ is an approximation of $(\alpha_k \rho_k)$ at time T^n. Here, we write the time discretization of the Cathare scheme:

$$
\begin{cases}
\frac{(\alpha_k \rho_k)^{n+1} - (\alpha_k \rho_k)^n}{\Delta t} + \partial_x (\alpha_k \rho_k u_k)^{n+1} = 0 \\[2mm]
(\alpha_k \rho_k)^{n+1} \frac{u_k^{n+1} - u_k^n}{\Delta t} + (\alpha_k \rho_k u_k)^{n+1} \partial_x u_k^{n+1} + \alpha_k^{n+1} \partial_x p^{n+1} = (\alpha_k \rho_k)^{n+1} g + F_k^{n,n+1} \\[2mm]
\frac{1}{\Delta t} \left[(\alpha_k \rho_k)^{n+1} \left(H_k + \frac{u_k^2}{2} \right)^{n,n+1} - (\alpha_k \rho_k)^n \left(H_k + \frac{u_k^2}{2} \right)^{n-1,n} \right] \\[2mm]
+ \partial_x \left[\alpha_k \rho_k u_k \left(H_k + \frac{u_k^2}{2} \right) \right]^{n+1} = \alpha_k^{n+1} \frac{p^{n+1} - p^n}{\Delta t} + (\alpha_k \rho_k u_k)^{n+1} g
\end{cases}
\tag{2}
$$

Where the notation $F_k^{n,n+1}$ means that the discretization of F_k is a function of the approximate solution at times T^n and T^{n+1}. After discretization, the non linear system is solved by a Newton method. Here, we highlight some characteristics of the Cathare scheme, some advantages and limitations:

- The scheme must be accurate enough at the incompressible limit to be able to capture the correct streamlines and pressure fields. This characteristic was mainly studied in the monophasic case:

  - Staggered schemes enjoy good precision at the incompressible limit.
  - However, Riemann solvers have poor precision in the incompressible limit. Corrections are proposed in [6] to overcome this issue.

- The vanishing phase is a numerical challenge of two phase flows simulation. It is important to capture well the volume fraction since it governs the composition of the mixture and two phase/single phase transition. An important issue is to guarantee the positivity of the volume fraction. Many schemes were designed to ensure this property (like [16] for two incompressible phases). Cathare uses a high interfacial friction to deal numerically with these transitions.

3 The Parareal Algorithm

Several approaches have been proposed over the years to decompose the time direction when solving a partial differential equation (see [9] for an overview). Of these, the parareal in time algorithm, which performances we explore in this work, has received an increasing amount of attention in the last 20 years with many applications (see [3, 8, 17] among many others). In the sequel, we recall the classical parareal algorithm as initially proposed in [3, 5, 13] and the principle of the multistep variant we will apply in Sect. 4.

3.1 Original Parareal Algorithm

After the discretization in space of a PDE with $\mathcal{N}$ the number of degrees of freedom:

$$\frac{\partial u}{\partial t} + A(t, u) = 0, \quad t \in [0, T], \quad u(t = 0) = u_0 \tag{3}$$

$$A : \mathbb{R} \times \mathbb{R}^{\mathcal{N}} \to \mathbb{R}^{\mathcal{N}}, u \in \mathbb{R}^{\mathcal{N}} \tag{4}$$

We recall here the classical parareal algorithm as initially proposed in [3, 5, 13]. Let G and F be two propagators such that, for any given $t \in [0, T]$, $s \in [0, T - t]$ and any function w in a Banach space, $G(t, s, w)$ (respectively $F(t, s, w)$) takes w as an initial value at time t and propagates it at time $t + s$. The full time interval is divided into N^c sub-intervals $[T^n, T^{n+1}]$ of size ΔT that will each be assigned to a processor. The algorithm is defined using two propagation operators:

- $G(T^n, \Delta T, u^n)$ computes a coarse approximation of $u(T^{n+1})$ with initial condition $u(T^n) \simeq u^n$ (low computational cost).
- $F(T^n, \Delta T, u^n)$ computes a more accurate approximation of $u(T^{n+1})$ with initial condition $u(T^n) \simeq u^n$ (high computational cost).

Starting from a coarse approximation u_0^n at times $T^0, T^1, \cdots, T^{N^c}$, obtained using G, the parareal algorithm performs for $k = 0, 1, \cdots$ the following iteration:

$$u_{k+1}^{n+1} = G(T^n, \Delta T, u_{k+1}^n) + F(T^n, \Delta T, u_k^n) - G(T^n, \Delta T, u_k^n)$$

In the parareal algorithm, the value $u(T^n)$ is approximated by u_k^n at each iteration k with an accuracy that tends rapidly to the one achieved by the fine solver, when k increases. The coarse approximation G can be chosen much less expensive than the fine solver F by the use of a scheme with a much larger time step (even $\delta T = \Delta T$) $\delta T \gg \delta t$ (time step of the fine solver) or by using a reduced model. All the fine propagations are made in parallel over the time windows and the coarse propagations are computed in a sequential way but have a low computational cost. The main convergence properties were studied in [10] and stability analysis was made in [4, 18]. We refer to [14] about the parallel efficiency of parareal and a recent work offering a new formulation of the algorithm to improve the parallel efficiency of the original one.

3.2   Multistep Variant of Parareal

This variant of the parareal algorithm was proposed in [1]. In the sequel, we will consider that the fine solver is based on a two step time scheme like the Cathare time scheme. Hence we will use the following notation for the fine solver that takes two initial values: $F(t, s, x, y)$, for $t \in [0, T]$, $s \in [0, T - t[$ and x, y in a Banach space.

Example If one solves (3) with a multistep time scheme as fine propagator F like the order 2 BDF method:

$$\frac{3}{2}u^{j+1} - 2u^j + \frac{1}{2}u^{j-1} = -\delta t A(u^{j+1}, t^{j+1}), \quad j = 1, \cdots, N^f, t^{j+1} - t^j = \delta t$$

Here the fine solver reads: $u^{j+1} = F(t^j, \delta t, u^{j-1}, u^j)$. Now, we apply the parareal algorithm with a coarse grid: $T^0, \cdots, T^{N^c}$ where: $T^{n+1} - T^n = \Delta T = N^f \delta t$.

Then we can write: $u(T^n + j\delta t) \simeq u^{n,j}, j = 1, \cdots, N^f, n = 1, \cdots, N^c$.

In order to perform the fine propagation, in a given time window $[T^n, T^{n+1}]$, we only need the local initial condition u_k^n and a consistent approximation of $u(T^n - \delta t)$.

In [2], the authors propose a consistent approximation in the context of the simulation of molecular dynamics. The proposed method was linked to the nature of the model and the symplectic character of their algorithm is shown, which is an important property to verify for molecular dynamics.

In the context of our application to the thermalhydraulic code Cathare, we want to derive a multistep variant of parareal that will not be intrusive in the software.

We seek a consistent approximation of $u(T^n - \delta t)$. The only fine trajectory at our disposal is $F(T^{n-1}, \Delta T, u_k^{n-2,N^f-1}, u_k^{n-1})$. Its final value at T^n is:

$F(T^{n-1}, \Delta T, u_k^{n-2,N^f-1}, u_k^{n-1})(T^n)$ from which we compute u_{k+1}^n by the parareal correction. Hence, we translate the solution:

$F(T^{n-1}, \Delta T - \delta t, u_k^{n-2,N^f-1}, u_k^{n-1})(T^n - \delta t)$ by the same correction:

$u_{k+1}^n - F(T^{n-1}, \Delta T, u_k^{n-2,N^f-1}, u_k^{n-1})$ and obtain the so called consistent approximation u_{k+1}^{n-1,N^f-1} to initialize the fine propagation in $[T^n, T^{n+1}]$. We now detail our algorithm:

$$
\begin{cases}
u_0^{n+1} = G(T^n, \Delta T, u_0^n), \quad 0 \le n \le N - 1 \\
u_{k+1}^{n+1} = G(T^n, \Delta T, u_{k+1}^n) + F(T^n, \Delta T, u_k^{n-1,N^f-1}, u_k^n) \\
\qquad - G(T^n, \Delta T, u_k^n), \quad 0 \le n \le N - 1, \quad k \ge 0 \\
u_{k+1}^{n,N^f-1} = F(T^n, \Delta T - \delta t, u_k^{n-1,N^f-1}, u_k^n) + u_{k+1}^{n+1} \\
\qquad - F(T^n, \Delta T, u_k^{n-1,N^f-1}, u_k^n), \quad 0 \le n \le N - 1, \quad k \ge 0
\end{cases}
\tag{5}
$$

Remark In order to perform the fine propagation, in a given time window $[T^n, T^{n+1}]$, at the first parareal iteration we need to choose a different consistent approximation of $u(T^n - \delta t)$, since we have not used the fine solver yet. To treat this, we could make one iteration with a Backward Euler method or one iteration with a second order Runge Kutta method. In the context of the application to the Cathare code, we choose a non intrusive initialization by imposing $u_0^{n-1,N^f-1} = u_0^n$.

4 Test Case

Here we apply the multistep parareal algorithm to the resolution of an oscillating manometer. This test case is proposed in [11] for system codes to test the ability of each numerical scheme to preserve system mass and to retain the gas-liquid interface. In this test case, the phases are separated and the interfacial friction term will be important in this configuration.

Note that here we have used the same physical model and the same mesh (110 cells) for both the coarse and the fine solvers: the only difference is the size of the time steps, δt for F and ΔT for G. All calculations have been evaluated with a stopping criteria where the tolerance is fixed to the precision of the numerical scheme, $\epsilon = 5 \cdot 10^{-2}$. With this threshold, parareal convergence is achieved after 2 or 3 iterations.

In the following subsections, after giving a numerical proof of the convergence of the parareal algorithm in our test case, some results about measured speed-up will be presented.

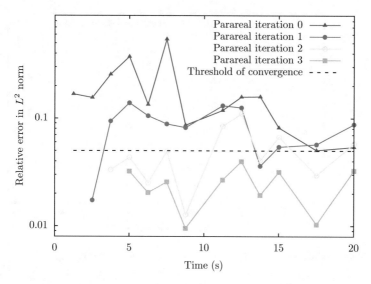

Fig. 1 Convergence of the multistep parareal algorithm when $\delta t = 10^{-5}$ and $\Delta T = 10\delta t$

4.1 About the Convergence

Figure 1 illustrates that the multistep parareal algorithm effectively converges when applied to the problem of the oscillating manometer. For a given time step T^n and parareal iteration k, the relative error in L^2 norm between the parareal solution and the sequential fine solver decreases beyond our given convergence threshold ϵ. In the figure, the test case has been solved with the multistep parareal algorithm when $\delta t = 10^{-5}$ and $\Delta T = 10\delta t$.

These results are obtained on 16 time windows and we will use this configuration in the sequel to study the performances.

4.2 Speed-Up Performances

In the following strong scaling tests, the same setting is used for the multistep parareal algorithm. The test case has been solved on an increasing number N_{proc} of processes $N_{proc} = 5, 10, 15, \cdots, 70$. In Fig. 2, with 25 processes, we obtain a speed up of 3.4 and of 3.7 with 50 processes. Here, we observe two global trends:

- For $N_{proc} = \{5, 10, 15, 20, 25, 40, 50\}$, the speed up first monotonically increases until reaching 25 processes and then increase again with 40 and 50 processes. This is due to the number of parareal iterations that is equal to 2 in this case.

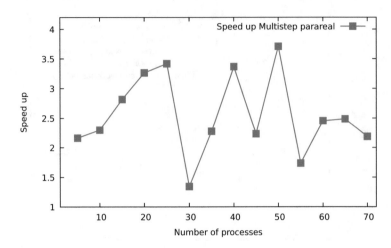

Fig. 2 Strong scaling results with the multistep variant of the parareal algorithm

- For $N_{proc} = \{30, 35, 45, 55, 60, 65, 70\}$, the speed up is drastically reduced because the parareal algorithm converges in three iterations in this case.

In the sequel, we highlight the well-known dependance of the computational cost of the parareal algorithm on the number of iterations. Let T_{fine} be the cpu time to run the fine solver in a sequential way on the whole time interval $[0, T]$. Since the coarse time step is ten times greater than the fine time step, we suppose that the cpu time of the coarse solver $T_{coarse} = \frac{T_{fine}}{10}$. This ratio between coarse and fine solvers should be as high as possible to minimize the computational cost of the coarse solver which is launched in a sequential way. This aspect will be studied in a forthcoming work to obtain better speed-up performances by coarsening more the solver G. When the algorithm converges in N_{it} iterations, the coarse solver is launched N_{it} times and the fine solver $N_{it} - 1$ times in parallel over the number of processes N_{proc}. Hence, we can write the cpu time in parallel T_{para} in terms of T_{fine}:

$$T_{para} = (N_{it} - 1)\frac{T_{fine}}{N_{proc}} + N_{it}T_{coarse} + \tau = \left(\frac{N_{it} - 1}{N_{proc}} + \frac{N_{it}}{10}\right) T_{fine} + \tau$$

where τ contains the time of communication between processes and the cpu time for the computation of the parareal corrections and of the error. Now, we can deduce an upper bound of the speed up S when the parareal algorithm converges in 2 or 3 iterations by neglecting τ:

$$S = \frac{T_{fine}}{T_{para}} \leq \frac{1}{\frac{N_{it}-1}{N_{proc}} + \frac{N_{it}}{10}}$$

Example On 25 processes, the algorithm converges in 2 iterations: $S \approx 4$ when the measured speed up is 3.4.

On 35 processes, the algorithm converges in 3 iterations: $S \approx 2.8$ when the measured speed up is 2.3.

5 Conclusion

The results of this study show that the parareal algorithm can effectively speed-up two phase flows simulations. Here we have tested the multistep variant of the parareal algorithm on a test case that is representative of the numerical challenges for two phase flows. These results can certainly be improved by using a new version of parareal (see [14]) that improves the parallel efficiency of the original parareal. This new algorithm proposes to solve the subproblems at increasing accuracy across the parareal iterations and should allow us to increase the speed up performances obtained on our test case.

Acknowledgments This research is partially supported by ANR project CINE-PARA (ANR-15-CE23-0019). We acknowledge the travel support provided by SMAI to attend the Spanish-French School Jacques-Louis Lions.

References

1. Ait-Ameur, K., Maday, Y., Tajchman, M.: Multi-step variant of the parareal algorithm. In: Haynes, R.D., MacLachlan, S., Cai, X.C., Halpern, L., Kim, H.H., Klawonn, A., Widlund, O. (eds.) Domain Decomposition Methods in Science and Engineering XXV, Lecture Notes in Computational Science and Engineering, pp. 393–400 (2020)
2. Audouze, C., Massot, M., Volz, S.: Symplectic multi-time step parareal algorithms applied to molecular dynamics. http://hal.archives-ouvertes.fr/hal-00358459/fr/ (2009)
3. Baffico, L., Bernard, S., Maday, Y., Turinici, G., Zrah, G.: Parallel-in-time molecular dynamics simulations. Phys. Rev. E **66**, 057701 (2002)
4. Bal, G.: Parallelization in time of (stochastic) ordinary differential equations. http://www.columbia.edu/gb2030/PAPERS/paralleltime.pdf (2003)
5. Bal, G., Maday, Y.: A "parareal" time discretization for non-linear PDE's with application to the pricing of an American put. In: Recent Developments in Domain Decomposition Methods, vol. 23, pp. 189–202 (2002)
6. Dellacherie, S.: Analysis of Godunov type schemes applied to the compressible Euler system at low Mach number. J. Comput. Phys. **229**(4), 978–1016 (2010)
7. Drew, D.A., Passman, S.L.: Theory of Multicomponent Fluids. Springer, New York (1999)
8. Fischer, P.F., Hecht, F., Maday, Y.: A parareal in time semi-implicit approximation of the Navier-Stokes equations. In: Kornhuber, R., et al. (eds.) Domain Decomposition Methods in Science and Engineering. Lecture Notes in Computational Science and Engineering, vol. 40, pp. 433–440, Springer, Berlin (2004)

9. Gander, M.J.: 50 years of time parallel time integration. In: Carraro, T., Geiger, M., Krkel, S., Rannacher, R. (eds.) Multiple Shooting and Time Domain Decomposition Methods, pp. 69–114. Springer, Cham (2015)

10. Gander, M.J., Vandewalle, S.: Analysis of the parareal time-parallel time-integration method. SIAM J. Sci. Comput. **29**(2), 556–578 (2007)

11. Hewitt, G.F., Delhaye, J.M., Zuber, N.: Multiphase Science and Technology, vol. 6. Springer, New York (1991)

12. M. Ishii, Thermo-fluid Dynamic Theory of Two-phase Flow. Eyrolles, Paris (1975)

13. Lions, J.-L., Maday, Y., Turinici, G.: Résolution par un schéma en temps "pararéel". C. R. Acad. Sci. Paris **332**(7), 661–668 (2001)

14. Maday, Y., Mula, O.: An adaptive parareal algorithm. J. Comput. Appl. Math. (2020). https://arxiv.org/pdf/1909.08333.pdf

15. Ndjinga, M.: Influence of interfacial pressure on the hyperbolicity of the two-fluid model. C. R. Acad. Sci. Paris Ser. I **344**, 407–412 (2007)

16. Ndjinga, M., Nguyen, T.P.K., Chalons, C.: A 2×2 hyperbolic system modelling incompressible two phase flows: theory and numerics. Nonlinear Differ. Equ. Appl. **24**, 36 (2017). https://doi.org/10.1007/s00030-017-0458-6

17. Samaddar, D., Newman, D.E., Sanchez, R.: Parallelization in time of numerical simulations of fully-developed plasma turbulence using the parareal algorithm. J. Comput. Phys. **229**(18), 6558–6573 (2010)

18. Staff, G.A., Ronquist, E.M.: Stability of the Parareal algorithm, In: Domain Decomposition Methods in Science and Engineering, Lecture Notes in Computational Science and Engineering, vol. 40, pp. 449–456. Springer, Berlin (2005)

Modelling of Bedload Sediment Transport for Weak and Strong Regimes

C. Escalante, E. D. Fernández-Nieto, T. Morales de Luna, and G. Narbona-Reina

Abstract A two-layer shallow water type model is proposed to describe bedload sediment transport for strong and weak interactions between the fluid and the sediment. The key point falls into the definition of the friction law between the two layers, which is a generalization of those introduced in Fernández-Nieto et al. (https://doi.org/10.1051/m2an/2016018). Moreover, we prove formally that the two-layer model converges to a Saint-Venant-Exner system (SVE) including gravitational effects when the ratio between the hydrodynamic and morphodynamic time scales is small. The SVE with gravitational effects is a degenerated nonlinear parabolic system, whose numerical approximation can be very expensive from a computational point of view, see for example T. Morales de Luna et al. (https://doi.org/10.1007/s10915-010-9447-1). In this work, gravitational effects are introduced into the two-layer system without any parabolic term, so the proposed model may be a advantageous solution to solve bedload sediment transport problems.

Keywords Bedload · Saint Venant Exner · Sediment friction law · Two-layer model

C. Escalante
Dpto. de A.M., E. e I.O., y Matemática Aplicada, Universidad de Málaga, Málaga, Spain
e-mail: escalante@uma.es

E. D. Fernández-Nieto · G. Narbona-Reina (✉)
Dpto. Matemática Aplicada I. ETS Arquitectura, Universidad de Sevilla, Sevilla, Spain
e-mail: edofer@us.es; gnarbona@us.es

T. Morales de Luna
Dpto. de Matemáticas, Universidad de Córdoba, Córdoba, Spain
e-mail: tomas.morales@uco.es

© The Author(s), under exclusive license to Springer Nature Switzerland AG 2021
D. Greiner et al. (eds.), *Numerical Simulation in Physics and Engineering: Trends and Applications*, SEMA SIMAI Springer Series 24,
https://doi.org/10.1007/978-3-030-62543-6_6

1 Introduction

Our goal is to obtain a general model for bedload sediment transport that is valid in any regime, for strong and weak interactions between the fluid and sediment.

In most models a weak interaction between the sediment and the fluid is assumed. In this case Saint-Venant-Exner models [7] are usually considered (SVE in what follows). For the case of high bedload transport rate, two-layer shallow water type model are considered instead, see for example [18, 20, 21]. In this work we focus into the definition of a two-layer shallow water type model that can be applied in both situations.

For the case of uniform flows the thickness of the moving sediment layer can be predicted, because erosion and deposition rates are equal in those situations. This is a general hypothesis that is assumed when modeling weak bedload transport. The usual approach is to consider a coupled system consisting of a Shallow Water system for the hydrodynamical part combined with a morphodynamical part given by the so-called Exner equation. The whole system is known as Saint Venant Exner system [7]. Exner equation depends on the definition of the solid transport discharge. Different classical definitions can be found for the solid transport discharge, for instance the ones given by Meyer-Peter and Müller [14], Van Rijn's [23], Einstein [5], Nielsen [17], Fernández-Luque and Van Beek [8], Ashida and Michiue [1], Engelund and Fredsoe [6], Kalinske [12], Charru [4], etc. A generalization of these classical models was introduced in [10] where the morphodynamical component is deduced from a Reynolds equation and includes gravitational effects in the sediment layer. Classical models do not take into account in general such gravitational effects because in their derivation the hypothesis of nearly horizontal sediment bed is used (see for example [13]).

In general, classical definitions for solid transport discharge can be written as follows,

$$\frac{q_b}{Q} = \text{sgn}(\tau) \frac{k_1}{(1-\varphi)} \theta^{m_1} (\theta - k_2 \theta_c)_+^{m_2} \left(\sqrt{\theta} - k_3 \sqrt{\theta_c}\right)_+^{m_3}, \tag{1}$$

where Q represents the characteristic discharge, $Q = d_s \sqrt{g(1/r - 1)d_s}$, $r = \rho_1/\rho_2$ is the density ratio, ρ_1 being the fluid density and ρ_2 the density of the sediment particles; d_s the mean diameter of the sediment particles, and φ is the averaged porosity. The coefficients k_l and m_l, $l = 1, 2, 3$, are positive constants that depend on the model. We usually find $m_2 = 0$ or $m_3 = 0$, for example, Meyer-Peter and Müller model takes $m_3 = 0$ and Ashida and Michiue's model uses $m_2 = 0$.

The Shields stress, θ, is defined as the ratio between the agitating and the stabilizing forces, $\theta = |\tau|d_s^2/(g(\rho_2 - \rho_1)d_s^3)$, τ being the shear stress at the bottom. For example, for Manning's law, we have $\tau = \rho_1 g h_1 n^2 u_1 |u_1|/h_1^{4/3}$. Where h_1 and u_1 are the thickness and the velocity of the fluid layer, respectively, and n is the Manning coefficient.

Finally, θ_c is the critical Shields stress. The positive part, $(\cdot)_+$, in the definition implies that the solid transport discharge is null if $\theta \leq k\theta_c$ (with $k = k_2$ when $m_2 > 0$ and $k = \sqrt{k_3}$ when $m_3 > 0$). If the velocity of the fluid is zero, $u_1 = 0$, we have $\theta = 0 < k\theta_c$, and for any model that can be written under the structure (1) we obtain that $q_b = 0$, which means that there is no movement of the sediment layer. This is even true when the sediment layer interface is not horizontal which is a consequence of the fact that classical models do not take into account gravitational effects.

In order to introduce gravitational effects in classical models, Fowler et al. proposed in [11] (see also [16]) a modification of the Meyer-Peter and Müller formula that consists in replacing θ by θ_{eff}, where:

$$\theta_{\text{eff}} = |\text{sgn}(u_1)\theta - \vartheta \partial_x(b + h_2)|, \tag{2}$$

with

$$\vartheta = \frac{\theta_c}{\tan \delta}, \tag{3}$$

δ being the angle of repose of the sediment particles. The sediment surface is defined by $z = b + h_2$, where h_2 is the thickness of the sediment layer and b the topography function or bedrock layer. Then, θ_{eff} is defined in terms of the gradient of sediment surface.

In [10], a multi-scale analysis is performed taking into account that the velocity of the sediment layer is smaller than the one of the fluid layer. This leads to a shallow water type system for the fluid layer and a lubrication Reynolds equation for the sediment one. The model includes gravitational effects and the authors deduce that it can also be seen as a modification of classical models: θ is replaced by the proposed values $\theta_{\text{eff}}^{(L)}$ or $\theta_{\text{eff}}^{(Q)}$, depending on whether the friction law between the fluid and the sediment is linear or quadratic. In the case when h_2 is of order of d_s/ϑ, for a linear friction law, the definition of the effective shear stress proposed in [10] can be written as follows:

$$\theta_{\text{eff}}^{(L)} = \left| \text{sgn}(u_1)\theta - \vartheta \partial_x(b + h_2) - \vartheta \frac{\rho_1}{\rho_2 - \rho_1} \partial_x(b + h_1 + h_2) \right|. \tag{4}$$

Let us remark that if the water free surface is horizontal, the definition of $\theta_{\text{eff}}^{(L)}$ coincides with θ_{eff} (2), proposed by Fowler et al. in [11]. Otherwise, the main difference is that this definition for the effective shear stress takes into account not only the gradient of the sediment surface but also the gradient of the water free surface.

For the case of a quadratic friction law, although the definition is a combination of the same components, it is rather different. In this case we can write the effective

Shields parameter proposed in [10] as follows:

$$
\theta_{\text{eff}}^{(Q)} = \left| \text{sgn}(u_1)\sqrt{\theta} - \sqrt{\frac{\vartheta \rho_1}{\rho_2 - \rho_1} |\partial_x(\frac{\rho_1}{\rho_2} h_1 + h_2 + b)|} \, \text{sgn}\left(\partial_x(\frac{\rho_1}{\rho_2} h_1 + h_2 + b)\right) \right|^2.
$$

(5)

In the case of submerged bedload sediment transport, the drag term is defined by a quadratic friction law. Thus, we should consider an effective Shields stress given by $\theta_{\text{eff}}^{(Q)}$. Nevertheless, in the bibliography θ_{eff} (2) is usually considered, regardless the fact that θ_{eff} is an approximation of $\theta_{\text{eff}}^{(L)}$ which is deduced from a linear friction law. In any case, considering the definitions θ_{eff} (2), $\theta_{\text{eff}}^{(L)}$ (4), or $\theta_{\text{eff}}^{(Q)}$ (5), means that the corresponding SVE system with gravitational effects is a parabolic degenerated partial differential system with non linear diffusion. Moreover, the system cannot be written as combination of a hyperbolic part plus a diffusion term.

Let us remark that in the literature a linearized version can be found, where gravitational effects are included by considering a classical SVE model with an additional viscous term, see for example [15, 22] and references therein. The drawback of this approach is that the diffusive term should not be present in stationary situations, for instance when the velocity is not high enough and sediment slopes are under the one given by the repose angle. In such situations it is necessary to include some external criteria that controls whether the diffusion term is applied or not. This is not the case in definitions (4) or (5) where the effective Shields stress is automatically limited by the effect of the Coulomb friction angle.

In this work we propose a two-layer shallow water model for bedload transport. The model converges to a generalization of SVE model with gravitational effects for low transport regimes while being valid for higher transport regimes as well. Moreover, it has the advantage that the inclusion of gravitational effects does not imply to approximate any non-linear parabolic degenerated term, as for the case of SVE model with gravitational effects.

In the next section we propose the new two-layer Shallow Water model for bedload transport. We also show the formal convergence to the SVE model and the associated energy balance.

2 Proposed Model

We consider a domain with two immiscible layers corresponding to water (upper layer) and sediment (lower layer). The sediment layer is in turn decomposed into a moving layer of thickness h_m and a sediment layer that does not move of thickness h_f, adjacent to the fixed bottom. These thicknesses are not fixed because there is an exchange of sediment material between the layers. Particles are eroded from the lower sediment layer and come into motion in the upper sediment layer. Conversely, particles from the upper layer are deposited into the lower sediment layer and stop moving.

We propose a 2D shallow water model is obtained by averaging on the vertical direction the Navier-Stokes equations and taking into account suitable boundary conditions. In particular, at the free surface we impose kinematic boundary conditions and vanishing pressure; at the bottom a Coulomb friction law is considered. The friction between water and sediment is introduced through the term F at the water/sediment interface and the mass transference term in the internal sediment interface is denoted by T. The general notation for the water layer corresponds to the subindex 1 and for the sediment layer to the subindex 2. Thus, the water of layer has a thickness h_1 and moves with horizontal velocity u_1. The thickness of the total sediment layer is denoted by $h_2 = h_f + h_m$, and the moving sediment layer h_m flows with velocity u_m. The fixed bottom or bedrock is denoted by b. See Fig. 1 for a sketch of the domain.

Note that the velocity of the sediment layer is defined as $u_2 = u_m$ in the moving layer and $u_2 = 0$ in the static layer. We assume an hydrostatic pressure regime.

Then we propose the following two-layer shallow water model:

$$\partial_t h_1 + \nabla \cdot (h_1 u_1) = 0 \tag{6a}$$

$$\partial_t (h_1 u_1) + \nabla \cdot (h_1 u_1 \otimes u_1) + g h_1 \nabla_x (b + h_1 + h_2) = -F \tag{6b}$$

$$\partial_t h_2 + \nabla \cdot (h_m u_m) = 0 \tag{6c}$$

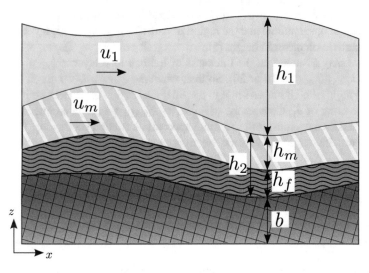

Fig. 1 Sketch of the domain for the fluid-sediment problem

$$\partial_t(h_m u_m) + \nabla \cdot (h_m u_m \otimes u_m) + gh_m \nabla_x(b + rh_1 + h_2)$$
$$= rF + \tfrac{1}{2}u_m T - (1 - r)gh_m \mathrm{sgn}(u_m)\tan\delta \tag{6d}$$

$$\partial_t h_f = -T \tag{6e}$$

where $r = \rho_1/\rho_2$ is the ratio between the densities of the water, ρ_1, and the sediment particles, ρ_2. δ is the internal Coulomb friction angle. In the next lines we give the closures for the friction term F and the mass transference T.

Following [10] we consider two types of friction laws at the interface: linear and quadratic. The friction term for the linear friction law is defined as

$$F_L = C_L(u_1 - u_m) \quad \text{with} \quad C_L = g\left(\frac{1}{r} - 1\right)\frac{h_1 h_m}{\vartheta(h_1 + h_m)\sqrt{(\frac{1}{r} - 1)gd_s}} \tag{7}$$

and for the quadratic friction law,

$$F_Q = C_Q(u_1 - u_m)|u_1 - u_m| \quad \text{with} \quad C_Q = \frac{1}{\beta}\frac{h_1 h_m}{\vartheta(h_1 + h_m)}, \tag{8}$$

d_s being the mean diameter of the sediment particles. ϑ is defined by Eq. (3). This definition of ϑ complies with the analysis of Seminara et al. [19], who concluded that the drag coefficient is proportional to $\tan(\delta)/\theta_c$.

Remark that the calibration coefficient β has units of length so that C_Q is non-dimensional. In [10], $\beta = d_s$ was assumed for the bedload in low transport situations. In our case, given that we deal with a complete bilayer system for any regime, this value is not always valid. In bedload framework, we can establish from experimental observations that the region of particles moving at this level is at most 10–20 particle-diameter in height [3].

So we may assume that the thickness of the bed load layer is $h_m = k\,d_s$ with $k \in [0, k_{\max}]$ ($k_{\max} = 10$ or 20). So that, when $h_m \leq k_{\max}d_s$ we are in a bedload low rate regime and it makes sense to consider the friction coefficient as in [10], that is, of the order of d_s. Conversely, when $h_m > k_{\max}d_s$ we are in an intense bedload regime and then we must turn to a more appropriate friction coefficient. Thus, to be consistent with our previous work, we propose to take:

$$\beta = \begin{cases} h_m & \text{if } h_m > k_{\max}d_s \\ d_s & \text{if } h_m \leq k_{\max}d_s \end{cases}$$

Another possibility would be to define $\beta = k_{\max}d_s$ when $h_m \leq k_{\max}d_s$. The coefficient $k_{\max}$ can be then considered as a calibration constant for the friction law.

The mass transference between the moving and the static sediment layers T is defined in terms of the difference between the erosion rate, $\dot{z}_e$, and the deposition rate $\dot{z}_d$. There exists in the literature different forms to close the definition of the erosion and deposition rates, all of them depending on calibration parameters (see for example [4]). For instance the following definitions are given in [9]:

$$T = \dot{z}_e - \dot{z}_d \quad \text{with} \quad \dot{z}_e = K_e(\theta_e - \theta_c) + \frac{\sqrt{g(1/r - 1)d_s}}{1 - \varphi}, \quad \dot{z}_d = K_d h_m \frac{\sqrt{g(1/r - 1)d_s}}{d_s}.$$

The coefficients K_e and K_d are erosion and deposition constants, respectively, φ is the porosity. For the case of nearly flat sediment bed, $\theta_e = \theta$ is usually set. This corresponds to the Bagnold's relation (see [2]). Nevertheless, in order to take into account the gradient of the sediment bed θ_e must be defined in terms of the effective Shields stress (see [10]). Then we define θ_e in terms of the friction law between the fluid and the sediment layers: for a linear friction law it is given by (4) and Eq. (5) gives its value for a quadratic friction law.

2.1 Convergence to the Classical SVE System for Weak Regimes

In this subsection we show formally the convergence of system (6) to the Saint-Venant-Exner model presented in [10]. This model is also obtained from an asymptotic approximation of the Navier-Stokes equations but following a different derivation for water and sediment under the hypothesis of large morphodynamic time scale. In particular, it has the following advantages: it preserves the mass conservation, the velocity (and hence, the discharge) of the bedload layer is explicitly deduced, and it has a dissipative energy balance.

The model introduced in [10] reads as follows:

$$\begin{cases} \partial_t h_1 + \nabla_x \cdot q_1 = 0, \\[2mm] \partial_t q_1 + \nabla_x \cdot \left(h_1(u_1 \otimes u_1)\right) + gh_1 \nabla_x(b + h_2 + h_1) = -\frac{gh_m}{r}\mathcal{P}, \\[2mm] \partial_t h_2 + \nabla_x \cdot \left(h_m v_b \sqrt{(1/r - 1)gd_s}\right) = 0, \\[2mm] \partial_t h_f = -T. \end{cases} \tag{9}$$

with

$$\mathcal{P} = \nabla_x(rh_1 + h_2 + b) + (1 - r)\mathrm{sgn}(u_2)\tan\delta. \tag{10}$$

The definition of the non-dimensional sediment velocity v_b depends on the friction law. When a linear friction law is considered, it reads:

$$v_b^{(LF)} = \frac{u_1}{\sqrt{(1/r - 1)gd_s}} - \frac{\vartheta}{1 - r}\mathcal{P},$$

(11)

where

$$\mathrm{sgn}(u_2) = \mathrm{sgn}\left(\frac{u_1}{\sqrt{(1/r - 1)gd_s}} - \frac{\vartheta}{1 - r}\nabla_x(rh_1 + h_2 + b)\right).$$

For a quadratic friction law:

$$v_b^{(QF)} = \frac{u_1}{\sqrt{(1/r - 1)gd_s}} - \left(\frac{\vartheta}{1 - r}\right)^{1/2}|\mathcal{P}|^{1/2}\mathrm{sgn}(\mathcal{P}),$$

(12)

where $\mathrm{sgn}(u_2) = \mathrm{sgn}(\Psi)$ and

$$\Psi = \frac{u_1}{\sqrt{(1/r - 1)gd_s}} - \left|\frac{\vartheta}{1 - r}\nabla_x(rh_1 + h_2 + b)\right|^{1/2}\mathrm{sgn}\left(\frac{\vartheta}{1 - r}\nabla_x(rh_1 + h_2 + b)\right).$$

The convergence is obtained when we assume the adequate asymptotic regime in terms of the time scales. As it is well known, for the weak bedload transport problem, the morphodynamic time is much larger than the hydrodynamic time, which makes the pressure effects much more important than the convective ones. As a consequence, the behavior of the sediment layer is just defined by the solid mass equation (Exner equation), omitting a momentum equation. This large morphodynamic time turns into an assumption of a smaller velocity for the lower layer. In order to fall into the low bedload transport regime we must also assume that the thickness of the bottom layer is smaller, because it represents the layer of moving sediment. Thus, we suppose:

$$u_m = \varepsilon_u \tilde{u}_m; \quad h_m = \varepsilon_h \tilde{h}_m; \quad T = \varepsilon_u \tilde{T}.$$

with ε_h and ε_u small parameters. Now we take these values into the momentum conservation equation for the lower layer (6d):

$$\partial_t(\varepsilon_h\varepsilon_u\tilde{h}_m\tilde{u}_m) + \nabla \cdot (\varepsilon_h\varepsilon_u^2\tilde{h}_m\tilde{u}_m \otimes \tilde{u}_m) + g\varepsilon_h\tilde{h}_m\nabla_x(b + rh_1 + h_2)$$
$$= r\tilde{F} + \varepsilon_u^2\frac{1}{2}\tilde{u}_m\tilde{T} - (1 - r)g\varepsilon_h\tilde{h}_m\mathrm{sgn}(\tilde{u}_m)\tan\delta$$

Then, if we neglect second order terms in $(\varepsilon_h, \varepsilon_u)$, we get

$$g\varepsilon_h\tilde{h}_m\nabla_x(b + rh_1 + h_2) = r\tilde{F} - (1 - r)g\varepsilon_h\tilde{h}_m\mathrm{sgn}(\tilde{u}_m)\tan\delta.$$

Returning to dimension variables, this equation reads:

$$r F = gh_m \nabla_x (b + rh_1 + h_2) + (1 - r)gh_m \mathrm{sgn}(u_m) \tan \delta = gh_m \mathcal{P},$$

where the last equality follows from the definition of $\mathcal{P}$, (10). Thus the expression of the friction term is

$$r F = gh_m \mathcal{P}; \tag{13}$$

which coincides with the friction term in the momentum equation of layer 1, r.h.s. of (9).

Now, from this equation and using the expressions of F, for linear (7) and quadratic (8) laws, we have to compute the value of u_m to check that it fits with (11) and (12) respectively.

- Linear friction law:

$$\tilde{F} = g\left(\frac{1}{r} - 1\right) \frac{1}{\vartheta \sqrt{(\frac{1}{r} - 1)gd_s}} \frac{\varepsilon_h \tilde{h}_m}{1 + \varepsilon_h \frac{\tilde{h}_m}{h_1}} (u_1 - \varepsilon_u \tilde{u}_m)$$

$$= g\left(\frac{1}{r} - 1\right) \frac{\varepsilon_h \tilde{h}_m}{\vartheta \sqrt{(\frac{1}{r} - 1)gd_s}} (u_1 - \varepsilon_u \tilde{u}_m) + O(\varepsilon_h^2)$$

where in the last equality we have used that $\dfrac{1}{1 + \varepsilon_h \frac{\tilde{h}_m}{h_1}} = 1 - \varepsilon_h \dfrac{\tilde{h}_m}{h_1} + O(\varepsilon_h^2)$.

So turning to the dimension variables and neglecting second order terms, Eq. (13) reads:

$$rg\left(\frac{1}{r} - 1\right) \frac{h_m}{\vartheta \sqrt{(\frac{1}{r} - 1)gd_s}} (u_1 - u_m) = gh_m \mathcal{P}.$$

From where we directly obtain that $u_m = v_b^{(LF)} \sqrt{(\frac{1}{r} - 1)gd_s}$.

- Quadratic friction law:

Note that in this case β reduces to d_s and then

$$\tilde{F} = \frac{\varepsilon_h h_1 \tilde{h}_m}{\vartheta d_s (h_1 + \varepsilon_h \tilde{h}_m)} (u_1 - \varepsilon_u \tilde{u}_m)|u_1 - \varepsilon_u \tilde{u}_m| = \frac{\varepsilon_h \tilde{h}_m}{\vartheta d_s} (u_1 - \varepsilon_u \tilde{u}_m)|u_1 - \varepsilon_u \tilde{u}_m| + O(\varepsilon_h^2).$$

Following the same reasoning as above, Eq. (13) reads:

$$r \frac{h_m}{\vartheta d_s} (u_1 - u_m)|u_1 - u_m| = gh_m \mathcal{P}.$$

From where we obtain that

$$r\frac{1}{\vartheta d_s}(u_1 - u_m)^2 = g\,\mathcal{P}\,\text{sgn}(\mathcal{P}) \quad \text{and then} \quad u_m = v_b^{(QF)}\sqrt{\left(\frac{1}{r} - 1\right)g d_s}.$$

2.2   Energy Balance

The proposed model has an exact dissipative energy balance, which is an easy consequence of two-layer shallow water systems. We obtain the following result.

Theorem 1 *System* (6) *admits a dissipative energy balance that reads:*

$$\partial_t \left(rh_1\frac{|u_1|^2}{2} + h_m\frac{|u_m|^2}{2} + \frac{1}{2}g(rh_1^2 + h_2^2) + g\,rh_1h_2 + gb(rh_1 + h_2) \right)$$

$$+\nabla \cdot \left(rh_1u_1\frac{|u_1|^2}{2} + h_mu_m\frac{|u_m|^2}{2} + g\,rh_1u_1(h_1 + h_2 + b) + gh_mu_m(rh_1 + h_2 + b) \right)$$

$$\leq -r(u_1 - u_m)F - (1 - r)gh_m|u_m|\tan\delta;$$

where the friction term F is given by (7) *or* (8).

The proof of the previous result is straightforward and for the sake of brevity we omit it.

Notice that classical SVE model does not verify in general a dissipative energy balance. In [10] a modification of a classical SVE models by including gravitational effects has been proposed that allows to verify this property. Nevertheless the proposed model in this work presents several advantages: it preserves the mass conservation, it accounts with a direct energy balance and it has a better structure to be solved from a numerical point of view. Moreover it can be applied for both regimes, weak and strong bedload transport without any a priori prescription. Numerical approximation and tests will be presented in a forthcoming paper.

Acknowledgments This research has been partially supported by the Spanish Government and FEDER through the coordinated Research projects MTM 2015-70490-C2-1-R and MTM 2015-70490-C2-2-R. The authors would like to thank M.J. Castro Díaz and R. Maurin for fruitful discussions.

References

1. Ashida, K., Michiue, M.: Study on hydraulic resistance and bedload transport rate in alluvial streams. JSCE Tokyo 206, 59–69 (1972)
2. Bagnold, R.A.: The flow of cohesionless grains in fluids. R. Soc. Lond. Philos. Trans. Ser. A. Math. Phys. Sci. **249**(964), 235–297 (1956)

3. Chanson, H.: The Hydraulics of Open Channel Flow: An Introduction. Elsevier Butterworth-Heinemann, Oxford (2004)
4. Charru, F.: Selection of the ripple length on a granular bed sheared by a liquid flow. Phys. Fluids 18, 121508 (2006)
5. Einstein, H.A.: Formulas for the transportation of bed load. ASCE **107**, 561–575 (1942)
6. Engelund, F., Dresoe, J.: A sediment transport model for straight alluvial channels. Nordic Hydrol. **7**, 293–306 (1976)
7. Exner, F.: Über die wechselwirkung zwischen wasser und geschiebe in flüssen. Sitzungsber. Akad. Wissenschaften pt. IIa. Bd. 134 (1925)
8. Fernández Luque, R., Van Beek, R.: Erosion and transport of bedload sediment. J. Hydraulaul. Res. **14**, 127–144 (1976)
9. Fernández-Nieto, E.D., Lucas, C., Morales de Luna, T., Cordier, S.: On the influence of the thickness of the sediment moving layer in the definition of the bedload trasport formula in Exner systems. Comp. Fluids **91**, 87–106 (2014)
10. Fernández-Nieto, E.D., Morales de Luna, T., Narbona-Reina, G., Zabsonré, J.D.: Formal deduction of the Saint-Venant-Exner model including arbitrarily sloping sediment beds and associated energy. ESAIM: M2AN **51**, 115–145 (2017)
11. Fowler, A.C., Kopteva, N., Oakley, C.: The formation of river channel. SIAM J. Appl. Math. **67**, 1016–1040 (2007)
12. Kalinske, A.A.: Criteria for determining sand transport by surface creep and saltation. Trans. AGU. **23**(2), 639–643 (1942)
13. Kovacs, A., Parker, G.: A new vectorial bedload formulation and its application to the time evolution of straight river channels. J. Fluid Mech. 267, 153–183 (1994)
14. Meyer-Peter, E., Müller, R.: Formulas for bedload transport. ASCE **107**, 561–575 (1942). Rep. 2nd Meet. Int. Assoc. Hydraul. Struct. Res., Stockolm, pp. 39–64
15. Michoski, C., Dawson, C., Mirabito, C., Kubatko, E.J., Wirasaet, D., Westerink, J.J.: Fully coupled methods for multiphase morphodynamics. Adv. Water Resour. **59**, 95–110 (2013)
16. Morales de Luna, T., Castro Díaz, M.J., Parés Madroñal, C.: A Duality Method for Sediment Transport Based on a Modified Meyer-Peter & Müller Model. J. Sci. Comp. **48**, 258–273 (2011)
17. Nielsen, P.: Coastal Bottom Boundary Layers and Sediment Transport. Advanced Series on Ocean Engineering, vol. 4. World Scientific Publishing, Singapore (1992)
18. Savary, C.: Transcritical transient flow over mobile bed. Two-layer shallow-water model. PhD thesis, Université catholique de Louvain (2007)
19. Seminara, G., Solari, L., Parker, G.: Bed load at low Shields stress on arbitrarily sloping beds: failure of the Bagnold hypothesis. Water Resour. Res. **38**, 11 (2002). https://doi.org/10.1029/2001WR000681
20. Spinewine, B.: Two-layer flow behaviour and the effects of granular dilatancy in dam-break induced sheet-flow. PhD Thesis n.76, Université catholique de Louvain (2005)
21. Swartenbroekx, C., Soares-Frazão, S., Spinewine, B., Guinot, V., Zech, Y.: Hyperbolicity preserving HLL solver for two-layer shallow-water equations applied to dam-break flows. In: Dittrich, A., Koll, Ka., Aberle, J., Geisenhainer, P. (eds.). River Flow 2010, vol. 2, pp. 1379–1387. Bundesanstalt fur Wasserbau (BAW), Karlsruhe (2010)
22. Tassi, P., Rhebergen, S., Vionnet, C., Bokhove, O.: A discontinuous Galerkin finite element model for morphodynamical evolution in shallow flows. Comp. Meth. App. Mech. Eng. **197**, 2930–2947 (2008)
23. Van Rijn, L.C.: Sediment transport (I): bed load transport. J. Hydraul. Div. Proc. ASCE **110**, 1431–1456 (1984)

Printed in the United States
by Baker & Taylor Publisher Services